Science 80

Teacher's Guide

CONTENTS

Author: **Alpha Omega Publications**
Editor: Alan Christopherson, M.S.

Alpha Omega Publications®

804 N. 2nd Ave. E., Rock Rapids, IA 51246-1759
© MCMXCVI by Alpha Omega Publications, Inc.   All rights reserved.
LIFEPAC is a registered trademark of Alpha Omega Publications, Inc.

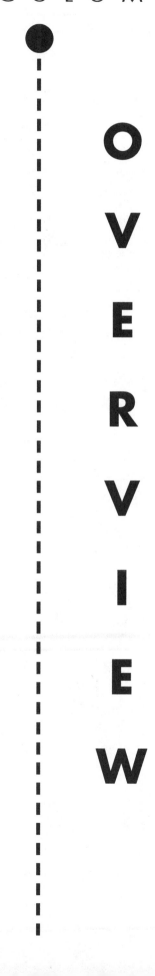

O
V
E
R
V
I
E
W

SCIENCE

Curriculum Overview
Grades 1–12

Science LIFEPAC Overview

	Grade 1	Grade 2	Grade 3
LIFEPAC 1	YOU LEARN WITH YOUR EYES • Name and group some colors • Name and group some shapes • Name and group some sizes • Help from what you see	THE LIVING AND NONLIVING • What God created • Rock and seed experiment • God-made objects • Man-made objects	YOU GROW AND CHANGE • Air we breathe • Food for the body • Exercise and rest • You are different
LIFEPAC 2	YOU LEARN WITH YOUR EARS • Sounds of nature and people • How sound moves • Sound with your voice • You make music	PLANTS • How are plants alike • Habitats of plants • Growth of plants • What plants need	PLANTS • Plant parts • Plant growth • Seeds and bulbs • Stems and roots
LIFEPAC 3	MORE ABOUT YOUR SENSES • Sense of smell • Sense of taste • Sense of touch • Learning with my senses	ANIMALS • How are animals alike • How are animals different • What animals need • Noah and the ark	ANIMAL AND ENVIRONMENT CHANGES • What changes an environment • How animals are different • How animals grow • How animals change
LIFEPAC 4	ANIMALS • What animals eat • Animals for food • Animals for work • Pets to care for	YOU • How are people alike • How are you different • Your family • Your health	YOU ARE WHAT YOU EAT • Food helps your body • Junk foods • Food groups • Good health habits
LIFEPAC 5	PLANTS • Big and small plants • Special plants • Plants for food • House plants	PET AND PLANT CARE • Learning about pets • Caring for pets • Learning about plants • Caring for plants	PROPERTIES OF MATTER • Robert Boyle • States of matter • Physical changes • Chemical changes
LIFEPAC 6	GROWING UP HEALTHY • How plants and animals grow • How your body grows • Eating and sleeping • Exercising	YOUR FIVE SENSES • Your eye • You can smell and hear • Your taste • You can feel	SOUNDS AND YOU • Making sounds • Different sounds • How sounds move • How sounds are heard
LIFEPAC 7	GOD'S BEAUTIFUL WORLD • Types of land • Water places • The weather • Seasons	PHYSICAL PROPERTIES • Colors • Shapes • Sizes • How things feel	TIMES AND SEASONS • The earth rotates • The earth revolves • Time changes • Seasons change
LIFEPAC 8	ALL ABOUT ENERGY • God gives energy • We use energy • Ways to make energy • Ways to save energy	OUR NEIGHBORHOOD • Things not living • Things living • Harm to our world • Caring for our world	ROCKS AND THEIR CHANGES • Forming rocks • Changing rocks • Rocks for buildings • Rock collecting
LIFEPAC 9	MACHINES AROUND YOU • Simple levers • Simple wheels • Inclined planes • Using machines	CHANGES IN OUR WORLD • Seasons • Change in plants • God's love never changes • God's Word never changes	HEAT ENERGY • Sources of heat • Heat energy • Moving heat • Benefits and problems of heat
LIFEPAC 10	WONDERFUL WORLD OF SCIENCE • Using your senses • Using your mind • You love yourself • You love the world	LOOKING AT OUR WORLD • Living things • Nonliving things • Caring for our world • Caring for ourselves	PHYSICAL CHANGES • Change in man • Change in plants • Matter and time • Sound and energy

Grade 4	Grade 5	Grade 6	
PLANTS • Plants and living things • Using plants • Parts of plants • The function of plants	**CELLS** • Cell composition • Plant and animal cells • Life of cells • Growth of cells	**PLANT SYSTEMS** • Parts of a plant • Systems of photosynthesis • Transport systems • Regulatory systems	LIFEPAC 1
ANIMALS • Animal structures • Animal behavior • Animal instincts • Man protects animals	**PLANTS: LIFE CYCLES** • Seed producing plants • Spore producing plants • One-celled plants • Classifying plants	**ANIMAL SYSTEMS** • Digestive system • Excretory system • Skeletal system • Diseases	LIFEPAC 2
MAN'S ENVIRONMENT • Resources • Balance in nature • Communities • Conservation and preservation	**ANIMALS: LIFE CYCLES** • Invertebrates • Vertebrates • Classifying animals • Relating function and structure	**PLANT AND ANIMAL BEHAVIOR** • Animal behavior • Plant behavior • Plant-animal interaction • Balance in nature	LIFEPAC 3
MACHINES • Work and energy • Simple machines • Simple machines together • Complex machines	**BALANCE IN NATURE** • Needs of life • Dependence on others • Prairie life • Stewardship of nature	**MOLECULAR GENETICS** • Reproduction • Inheritance • DNA and mutations • Mendel's work	LIFEPAC 4
ELECTRICITY AND MAGNETISM • Electric current • Electric circuits • Magnetic materials • Electricity and magnets	**TRANSFORMATION OF ENERGY** • Work and energy • Heat energy • Chemical energy • Energy sources	**CHEMICAL STRUCTURE** • Nature of matter • Periodic Table • Diagrams of atoms • Acids and bases	LIFEPAC 5
PROPERTIES OF MATTER • Properties of water • Properties of matter • Molecules and atoms • Elements	**RECORDS IN ROCK: THE FLOOD** • The Biblical account • Before the flood • The flood • After the flood	**LIGHT AND SOUND** • Sound waves • Light waves • The visible spectrum • Colors	LIFEPAC 6
WEATHER • Causes of weather • Forces of weather • Observing weather • Weather instruments	**RECORDS IN ROCK: FOSSILS** • Fossil types • Fossil location • Identifying fossils • Reading fossils	**MOTION AND ITS MEASUREMENT** • Definition of force • Rate of doing work • Laws of motion • Change in motion	LIFEPAC 7
THE SOLAR SYSTEM • Our solar system • The big universe • Sun and planets • Stars and space	**RECORDS IN ROCK: GEOLOGY** • Features of the earth • Rock of the earth • Forces of the earth • Changes in the earth	**SPACESHIP EARTH** • Shape of the earth • Rotation and revolution • Eclipses • The solar system	LIFEPAC 8
THE PLANET EARTH • The atmosphere • The hydrosphere • The lithosphere • Rotation and revolution	**CYCLES IN NATURE** • Properties of matter • Changes in matter • Natural cycles • God's order	**ASTRONOMY AND THE STARS** • History of astronomy • Investigating stars • Major stars • Constellations	LIFEPAC 9
GOD'S CREATION • Earth and solar system • Matter and weather • Using nature • Conservation	**LOOK AHEAD** • Plant and animal life • Balance in nature • Biblical records • Records of rock	**THE EARTH AND THE UNIVERSE** • Plant systems • Animal systems • Physics and chemistry • The earth and stars	LIFEPAC 10

Science LIFEPAC Overview

	Grade 7	Grade 8	Grade 9
LIFEPAC 1	**WHAT IS SCIENCE** • Tools of a scientist • Methods of a scientist • Work of a scientist • Careers in science	**SCIENCE AND SOCIETY** • Definition of science • History of science • Science today • Science tomorrow	**OUR ATOMIC WORLD** • Structure of matter • Radioactivity • Atomic nuclei • Nuclear energy
LIFEPAC 2	**PERCEIVING THINGS** • History of the metric system • Metric units • Advantages of the metric system • Graphing data	**STRUCTURE OF MATTER I** • Properties of matter • Chemical properties of matter • Atoms and molecules • Elements, compounds, & mixtures	**VOLUME, MASS, AND DENSITY** • Measure of matter • Volume • Mass • Density
LIFEPAC 3	**EARTH IN SPACE I** • Ancient stargazing • Geocentric Theory • Copernicus • Tools of astronomy	**STRUCTURE OF MATTER II** • Changes in matter • Acids • Bases • Salts	**PHYSICAL GEOLOGY** • Earth structures • Weathering and erosion • Sedimentation • Earth movements
LIFEPAC 4	**EARTH IN SPACE II** • Solar energy • Planets of the sun • The moon • Eclipses	**HEALTH AND NUTRITION** • Foods and digestion • Diet • Nutritional diseases • Hygiene	**HISTORICAL GEOLOGY** • Sedimentary rock • Fossils • Crustal changes • Measuring time
LIFEPAC 5	**THE ATMOSPHERE** • Layers of the atmosphere • Solar effects • Natural cycles • Protecting the atmosphere	**ENERGY I** • Kinetic and potential energy • Other forms of energy • Energy conversions • Entropy	**BODY HEALTH I** • Microorganisms • Bacterial infections • Viral infections • Other infections
LIFEPAC 6	**WEATHER** • Elements of weather • Air masses and clouds • Fronts and storms • Weather forecasting	**ENERGY II** • Magnetism • Current and static electricity • Using electricity • Energy sources	**BODY HEALTH II** • Body defense mechanisms • Treating disease • Preventing disease • Community health
LIFEPAC 7	**CLIMATE** • Climate and weather • Worldwide climate • Regional climate • Local climate	**MACHINES I** • Measuring distance • Force • Laws of Newton • Work	**ASTRONOMY** • Extent of the universe • Constellations • Telescopes • Space explorations
LIFEPAC 8	**HUMAN ANATOMY I** • Cell structure and function • Skeletal and muscle systems • Skin • Nervous system	**MACHINES II** • Friction • Levers • Wheels and axles • Inclined planes	**OCEANOGRAPHY** • History of oceanography • Research techniques • Geology of the ocean • Properties of the ocean
LIFEPAC 9	**HUMAN ANATOMY II** • Respiratory system • Circulatory system • Digestive system • Endocrine system	**BALANCE IN NATURE** • Photosynthesis • Food • Natural cycles • Balance in nature	**SCIENCE AND TOMORROW** • The land • Waste and ecology • Industry and energy • New frontiers
LIFEPAC 10	**CAREERS IN SCIENCE** • Scientists at work • Astronomy • Meteorology • Medicine	**SCIENCE AND TECHNOLOGY** • Basic science • Physical science • Life science • Vocations in science	**SCIENTIFIC APPLICATIONS** • Measurement • Practical health • Geology and astronomy • Solving problems

Grade 10	Grade 11	Grade 12
TAXONOMY • History of taxonomy • Binomial nomenclature • Classification • Taxonomy	**INTRODUCTION TO CHEMISTRY** • Metric units and instrumentation • Observation and hypothesizing • Scientific notation • Careers in chemistry	**KINEMATICS** • Scalars and vectors • Length measurement • Acceleration • Fields and models
BASIS OF LIFE • Elements and molecules • Properties of compounds • Chemical reactions • Organic compounds	**BASIC CHEMICAL UNITS** • Alchemy • Elements • Compounds • Mixtures	**DYNAMICS** • Newton's Laws of Motion • Gravity • Circular motion • Kepler's Laws of Motion
MICROBIOLOGY • The microscope • Protozoan • Algae • Microorganisms	**GASES AND MOLES** • Kinetic theory • Gas laws • Combined gas law • Moles	**WORK AND ENERGY** • Mechanical energy • Conservation of energy • Power and efficiency • Heat energy
CELLS • Cell theories • Examination of the cell • Cell design • Cells in organisms	**ATOMIC MODELS** • Historical models • Modern atomic structure • Periodic Law • Nuclear reactions	**WAVES** • Energy transfers • Reflection and refraction of waves • Diffraction and interference • Sound waves
PLANTS: GREEN FACTORIES • The plant cell • Anatomy of the plant • Growth and function of plants • Plants and people	**CHEMICAL FORMULAS** • Ionic charges • Electronegativity • Chemical bonds • Molecular shape	**LIGHT** • Speed of light • Mirrors • Lenses • Models of light
HUMAN ANATOMY AND PHYSIOLOGY • Digestive and excretory system • Respiratory and circulatory system • Skeletal and muscular system • Body control systems	**CHEMICAL REACTIONS** • Detecting reactions • Energy changes • Reaction rates • Equilibriums	**STATIC ELECTRICITY** • Nature of charges • Transfer of charges • Electric fields • Electric potential
INHERITANCE • Gregor Mendel's experiments • Chromosomes and heredity • Molecular genetics • Human genetics	**EQUILIBRIUM SYSTEMS** • Solutions • Solubility equilibriums • Acid-base equilibriums • Redox equilibriums	**CURRENT ELECTRICITY** • Electromotive force • Electron flow • Resistance • Circuits
CELL DIVISION & REPRODUCTION • Mitosis and meiosis • Asexual reproduction • Sexual reproduction • Plant reproduction	**HYDROCARBONS** • Organic compounds • Carbon atoms • Carbon bonds • Saturated and unsaturated	**MAGNETISM** • Fields • Forces • Electromagnetism • Electron beams
ECOLOGY & ENERGY • Ecosystems • Communities and habitats • Pollution • Energy	**CARBON CHEMISTRY** • Saturated and unsaturated • Reaction types • Oxygen groups • Nitrogen groups	**ATOMIC AND NUCLEAR PHYSICS** • Electromagnetic radiation • Quantum theory • Nuclear theory • Nuclear reaction
APPLICATIONS OF BIOLOGY • Principles of experimentation • Principles of reproduction • Principles of life • Principles of ecology	**ATOMS TO HYDROCARBONS** • Atoms and molecules • Chemical bonding • Chemical systems • Organic chemistry	**KINEMATICS TO NUCLEAR PHYSICS** • Mechanics • Wave motion • Electricity • Modern physics

LIFEPAC 1 · LIFEPAC 2 · LIFEPAC 3 · LIFEPAC 4 · LIFEPAC 5 · LIFEPAC 6 · LIFEPAC 7 · LIFEPAC 8 · LIFEPAC 9 · LIFEPAC 10

MANAGEMENT

STRUCTURE OF THE LIFEPAC CURRICULUM

The LIFEPAC curriculum is conveniently structured to provide one teacher handbook containing teacher support material with answer keys and ten student worktexts for each subject at grade levels two through twelve. The worktext format of the LIFEPACs allows the student to read the textual information and complete workbook activities all in the same booklet. The easy to follow LIFEPAC numbering system lists the grade as the first number(s) and the last two digits as the number of the series. For example, the Language Arts LIFEPAC at the 6th grade level, 5th book in the series would be LAN0605.

Each LIFEPAC is divided into 3 to 5 sections and begins with an introduction or overview of the booklet as well as a series of specific learning objectives to give a purpose to the study of the LIFEPAC. The introduction and objectives are followed by a vocabulary section which may be found at the beginning of each section at the lower levels, at the beginning of the LIFEPAC in the middle grades, or in the glossary at the high school level. Vocabulary words are used to develop word recognition and should not be confused with the spelling words introduced later in the LIFEPAC. The student should learn all vocabulary words before working the LIFEPAC sections to improve comprehension, retention, and reading skills.

Each activity or written assignment has a number for easy identification, such as 1.1. The first number corresponds to the LIFEPAC section and the number to the right of the decimal is the number of the activity.

Teacher checkpoints, which are essential to maintain quality learning, are found at various locations throughout the LIFEPAC. The teacher should check 1) neatness of work and penmanship, 2) quality of understanding (tested with a short oral quiz), 3) thoroughness of answers (complete sentences and paragraphs, correct spelling, etc.), 4) completion of activities (no blank spaces), and 5) accuracy of answers as compared to the answer key (all answers correct).

The self test questions are also number coded for easy reference. For example, 2.015 means that this is the 15th question in the self test of Section II. The first number corresponds to the LIFEPAC section, the zero indicates that it is a self test question, and the number to the right of the zero the question number.

The LIFEPAC test is packaged at the centerfold of each LIFEPAC. It should be removed and put aside before giving the booklet to the student for study.

Answer and test keys have the same numbering system as the LIFEPACs and appear at the back of this handbook. The student may be given access to the answer keys (not the test keys) under teacher supervision so that he can score his own work.

A thorough study of the Curriculum Overview by the teacher before instruction begins is essential to the success of the student. The teacher should become familiar with expected skill mastery and understand how these grade level skills fit into the overall skill development of the curriculum. The teacher should also preview the objectives that appear at the beginning of each LIFEPAC for additional preparation and planning.

TEST SCORING and GRADING

Answer keys and test keys give examples of correct answers. They convey the idea, but the student may use many ways to express a correct answer. The teacher should check for the essence of the answer, not for the exact wording. Many questions are high level and require thinking and creativity on the part of the student. Each answer should be scored based on whether or not the main idea written by the student matches the model example. "Any Order" or "Either Order" in a key indicates that no particular order is necessary to be correct.

Most self tests and LIFEPAC tests at the lower elementary levels are scored at 1 point per question; however, the upper levels may have a point system awarding 2 to 5 points for various questions. Further, the total test points will vary; they may not always equal 100 points. They may be 78, 85, 100, 105, etc.

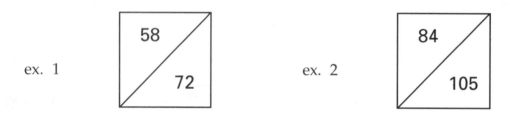

A score box similar to ex.1 above is located at the end of each self test and on the front of the LIFEPAC test. The bottom score, 72, represents the total number of points possible on the test. The upper score, 58, represents the number of points your student will need to receive an 80% or passing grade. If you wish to establish the exact percentage that your student has achieved, find the total points of his correct answers and divide it by the bottom number (in this case 72.) For example, if your student has a point total of 65, divide 65 by 72 for a grade of 90%. Referring to ex. 2, on a test with a total of 105 possible points, the student would have to receive a minimum of 84 correct points for an 80% or passing grade. If your student has received 93 points, simply divide the 93 by 105 for a percentage grade of 89%. Students who receive a score below 80% should review the LIFEPAC and retest using the appropriate Alternate Test found in the Teacher's Guide.

The following is a guideline to assign letter grades for completed LIFEPACs based on a maximum total score of 100 points.

LIFEPAC Test = 60% of the Total Score (or percent grade)
Self Test = 25% of the Total Score (average percent of self tests)
Reports = 10% or 10* points per LIFEPAC
Oral Work = 5% or 5* points per LIFEPAC
*Determined by the teacher's subjective evaluation of the student's daily work.

Example:

LIFEPAC Test Score	=	92%	92	x	.60	=	55 points
Self Test Average	=	90%	90	x	.25	=	23 points
Reports						=	8 points
Oral Work						=	4 points

TOTAL POINTS   =   90 points

Grade Scale based on point system:

100	–	94	=	A
93	–	86	=	B
85	–	77	=	C
76	–	70	=	D
Below		70	=	F

TEACHER HINTS and STUDYING TECHNIQUES

LIFEPAC Activities are written to check the level of understanding of the preceding text. The student may look back to the text as necessary to complete these activities; however, a student should never attempt to do the activities without reading (studying) the text first. Self tests and LIFEPAC tests are never open book tests.

Language arts activities (skill integration) often appear within other subject curriculum. The purpose is to give the student an opportunity to test his skill mastery outside of the context in which it was presented.

Writing complete answers (paragraphs) to some questions is an integral part of the LIFEPAC Curriculum in all subjects. This builds communication and organization skills, increases understanding and retention of ideas, and helps enforce good penmanship. Complete sentences should be encouraged for this type of activity. Obviously, single words or phrases do not meet the intent of the activity, since multiple lines are given for the response.

Review is essential to student success. Time invested in review where review is suggested will be time saved in correcting errors later. Self tests, unlike the section activities, are closed book. This procedure helps to identify weaknesses before they become too great to overcome. Certain objectives from self tests are cumulative and test previous sections; therefore, good preparation for a self test must include all material studied up to that testing point.

The following procedure checklist has been found to be successful in developing good study habits in the LIFEPAC curriculum.

1. Read the introduction and Table of Contents.
2. Read the objectives.
3. Recite and study the entire vocabulary (glossary) list.
4. Study each section as follows:
    a. Read the introduction and study the section objectives.
    b. Read all the text for the entire section, but answer none of the activities.
    c. Return to the beginning of the section and memorize each vocabulary word and definition.
    d. Reread the section, complete the activities, check the answers with the answer key, correct all errors, and have the teacher check.
    e. Read the self test but do not answer the questions.
    f. Go to the beginning of the first section and reread the text and answers to the activities up to the self test you have not yet done.
    g. Answer the questions to the self test without looking back.
    h. Have the self test checked by the teacher.
    i. Correct the self test and have the teacher check the corrections.
    j. Repeat steps a–i for each section.

5. Use the SQ3R* method to prepare for the LIFEPAC test.
6. Take the LIFEPAC test as a closed book test.
7. LIFEPAC tests are administered and scored under direct teacher supervision. Students who receive scores below 80% should review the LIFEPAC using the SQ3R* study method and take the Alternate Test located in the Teacher Handbook. The final test grade may be the grade on the Alternate Test or an average of the grades from the original LIFEPAC test and the Alternate Test.

    *SQ3R: Scan the whole LIFEPAC.

               Question yourself on the objectives.

               Read the whole LIFEPAC again.

               Recite through an oral examination.

               Review weak areas.

GOAL SETTING and SCHEDULES

Each school must develop its own schedule, because no single set of procedures will fit every situation. The following is an example of a daily schedule that includes the five LIFEPAC subjects as well as time slotted for special activities.

Possible Daily Schedule

8:15	–	8:25	Pledges, prayer, songs, devotions, etc.
8:25	–	9:10	Bible
9:10	–	9:55	Language Arts
9:55	–	10:15	Recess (juice break)
10:15	–	11:00	Mathematics
11:00	–	11:45	Social Studies
11:45	–	12:30	Lunch, recess, quiet time
12:30	–	1:15	Science
1:15	–		Drill, remedial work, enrichment*

*Enrichment: Computer time, physical education, field trips, fun reading, games and puzzles, family business, hobbies, resource persons, guests, crafts, creative work, electives, music appreciation, projects.

Basically, two factors need to be considered when assigning work to a student in the LIFEPAC curriculum.

The first is time. An average of 45 minutes should be devoted to each subject, each day. Remember, this is only an average. Because of extenuating circumstances a student may spend only 15 minutes on a subject one day and the next day spend 90 minutes on the same subject.

The second factor is the number of pages to be worked in each subject. A single LIFEPAC is designed to take 3 to 4 weeks to complete. Allowing about 3-4 days for LIFEPAC introduction, review, and tests, the student has approximately 15 days to complete the LIFEPAC pages. Simply take the number of pages in the LIFEPAC, divide it by 15 and you will have the number of pages that must be completed on a daily basis to keep the student on schedule. For example, a LIFEPAC containing 45 pages will require 3 completed pages per day. Again, this is only an average. While working a 45 page LIFEPAC, the student may complete only 1 page the first day if the text has a lot of activities or reports, but go on to complete 5 pages the next day.

Long range planning requires some organization. Because the traditional school year originates in the early fall of one year and continues to late spring of the following year, a calendar should be devised that covers this period of time. Approximate beginning and completion dates can be noted on the calendar as well as special occasions such as holidays, vacations and birthdays. Since each LIFEPAC takes 3-4 weeks or eighteen days to complete, it should take about 180 school days to finish a set of ten LIFEPACs. Starting at the beginning school date, mark off eighteen school days on the calendar and that will become the targeted completion date for the first LIFEPAC. Continue marking the calendar until you have established dates for the remaining nine LIFEPACs making adjustments for previously noted holidays and vacations. If all five subjects are being used, the ten established target dates should be the same for the LIFEPACs in each subject.

FORMS

The sample weekly lesson plan and student grading sheet forms are included in this section as teacher support materials and may be duplicated at the convenience of the teacher.

The student grading sheet is provided for those who desire to follow the suggested guidelines for assignment of letter grades found on page 3 of this section. The student's self test scores should be posted as percentage grades. When the LIFEPAC is completed the teacher should average the self test grades, multiply the average by .25 and post the points in the box marked self test points. The LIFEPAC percentage grade should be multiplied by .60 and posted. Next, the teacher should award and post points for written reports and oral work. A report may be any type of written work assigned to the student whether it is a LIFEPAC or additional learning activity. Oral work includes the student's ability to respond orally to questions which may or may not be related to LIFEPAC activities or any type of oral report assigned by the teacher. The points may then be totaled and a final grade entered along with the date that the LIFEPAC was completed.

The Student Record Book which was specifically designed for use with the Alpha Omega curriculum provides space to record weekly progress for one student over a nine week period as well as a place to post self test and LIFEPAC scores. The Student Record Books are available through the current Alpha Omega catalog; however, unlike the enclosed forms these books are not for duplication and should be purchased in sets of four to cover a full academic year.

WEEKLY LESSON PLANNER

Week of:

Subject	Subject	Subject	Subject

Monday

Subject	Subject	Subject	Subject

Tuesday

Subject	Subject	Subject	Subject

Wednesday

Subject	Subject	Subject	Subject

Thursday

Subject	Subject	Subject	Subject

Friday

WEEKLY LESSON PLANNER

Week of:

	Subject	Subject	Subject	Subject
Monday				
Tuesday				
Wednesday				
Thursday				
Friday				

Student Name _____   Year _____

Bible

LP #	Self Test Scores by Sections 1	2	3	4	5	Self Test Points	LIFEPAC Test	Oral Points	Report Points	Final Grade	Date
01											
02											
03											
04											
05											
06											
07											
08											
09											
10											

History & Geography

LP #	Self Test Scores by Sections 1	2	3	4	5	Self Test Points	LIFEPAC Test	Oral Points	Report Points	Final Grade	Date
01											
02											
03											
04											
05											
06											
07											
08											
09											
10											

Language Arts

LP #	Self Test Scores by Sections 1	2	3	4	5	Self Test Points	LIFEPAC Test	Oral Points	Report Points	Final Grade	Date
01											
02											
03											
04											
05											
06											
07											
08											
09											
10											

Student Name _____   Year _____

Mathematics

LP #	Self Test Scores by Sections 1	2	3	4	5	Self Test Points	LIFEPAC Test	Oral Points	Report Points	Final Grade	Date
01											
02											
03											
04											
05											
06											
07											
08											
09											
10											

Science

LP #	Self Test Scores by Sections 1	2	3	4	5	Self Test Points	LIFEPAC Test	Oral Points	Report Points	Final Grade	Date
01											
02											
03											
04											
05											
06											
07											
08											
09											
10											

Spelling/Electives

LP #	Self Test Scores by Sections 1	2	3	4	5	Self Test Points	LIFEPAC Test	Oral Points	Report Points	Final Grade	Date
01											
02											
03											
04											
05											
06											
07											
08											
09											
10											

N
O
T
E
S

INSTRUCTIONS FOR SCIENCE

The LIFEPAC curriculum from grades two through twelve is structured so that the daily instructional material is written directly into the LIFEPACs. The student is encouraged to read and follow this instructional material in order to develop independent study habits. The teacher should introduce the LIFEPAC to the student, set a required completion schedule, complete teacher checks, be available for questions regarding both content and procedures, administer and grade tests, and develop additional learning activities as desired. Teachers working with several students may schedule their time so that students are assigned to a quiet work activity when it is necessary to spend instructional time with one particular student.

The Teacher Notes section of the Teacher's Guide lists the required or suggested materials for the LIFEPACs and provides additional learning activities for the students. The materials section refers only to LIFEPAC materials and does not include materials which may be needed for the additional activities. Additional learning activities provide a change from the daily school routine, encourage the student's interest in learning and may be used as a reward for good study habits.

If you have limited facilities and are not able to perform all the experiments contained in the LIFEPAC curriculum, the Science Project List for grades 3-12 may be a useful tool for you. This list prioritizes experiments into three categories: those essential to perform, those which should be performed as time and facilities permit, and those not essential for mastery of LIFEPACs. Of course, for complete understanding of concepts and student participation in the curriculum, all experiments should be performed whenever practical. Materials for the experiments are shown in Teacher Notes – Materials Needed.

Science Projects List
Key

(1) = Those essential to perform for basic understanding of scientific principles.

(2) = Those which should be performed as time permits.

(3) = Those not essential for mastery of LIFEPACs.

S = Equipment needed for home school or Christian school lab.

E = Explanation or demonstration by instructor may replace student or class lab work.

H = Suitable for homework or for home school students. (No lab equipment needed.)

V = This experiment is available on the Science Experiments video.

Science 801
| pp | 16 | (1) | S & V |

Science 802
pp	4	(1)	H & V
	8	(1)	S
	10	(2)	E & V
	11	(1)	H & V
	13	(2)	H
	14	(2)	S
	29	(1)	S
	34	(2)	H
	37	(1)	H & V

Science 803
pp	4	(1)	H
	7	(1)	H
	8	(3)	E
	11	(1)	S
	13	(1)	S
	18	(1)	H
	29	(1)	H
	30	(1)	H
	37	(1)	S
	38	(1)	S
	43	(1)	S

Science 804
| pp | 8 | (1) | H |
| | 17 | (1) | H |

Science 805
None

Science 806
pp	3	(1)	S
	5	(1)	S
	7	(1)	S & V
	17	(1)	S & V
	34	(2)	H

Science 807
| pp | 16 | (1) | S |
| | 32 | (1) | S |

Science 808
pp	4	(1)	H
	5	(1)	H
	8	(1)	H
	11	(2)	H
	22	(1)	H or S
	25	(1)	H
	29	(2)	H
	34	(1)	H

Science 809
pp	4	(1)	S
	6	(1)	H
	43	(3)	E

Science 810
pp	9	(1)	S
	12	(2)	H
	16	(1)	S
	17	(3)	H
	18	(2)	S
	26	(2)	S
	30	(1)	H
	33	(1)	S
	41	(1)	H

Materials Needed for LIFEPAC:

Required:
Encyclopedia
ruler at least 10 centimeters long
graduated cylinder marked in
  milliliters
balance scale (triple beam or
  other type)

Suggested:
8th Grade Science Experiments video

Additional Learning Activities

Section I Science Today

1. Direct the student(s) to make a chart of events of science and technology in chronological order.
2. Use the charts to develop a time line of events in science and technology. This time line could be used also in the social sciences. Additional reference materials may be used to complete this activity.
3. Take a friend and a tape recorder and talk to someone who is over sixty years old. Ask questions about how the person lived when he or she was a child. What kind of medicine did the doctors have? Be certain to prepare a list of questions in advance.
4. Read a book on the history of science, one area of science, or one scientist.

Section II Science and Technology

1. With a friend develop a method to test the tensile strength of materials such as rubber bands, string, fine wire, etc. Test several items.
2. With friends make designs using potatoes. Slice the potatoes to make a flat surface. Cut in a design. Ink the potatoes with a stamp pad. Compare what can be done by this method with what can be done with Gutenberg's movable type.
3. Make squares one centimeter on each side on index cards. Spread the cards with petroleum jelly. Place the cards around school and home. Leave them for three days. Count the number of particles stuck to each square. Take the average of the cards. Where was the pollution greatest? Why?
4. Read the newspaper. Clip articles which relate to conflicts between science, technology, and society.
5. Design and build a model bridge. Test it to see how strong it is. Use straws, balsa, or toothpicks.

Section III Science and Technology of Tomorrow

1. From old magazines have students cut pictures of futuristic living. Explain how to make a collage. Have the students make a collage.
2. In the public library look up architecture. See changes that have taken place in buildings. Check names like Frank Lloyd Wright and Paoli Soleri.
3. From the encyclopedia or almanac get figures on the United States population for the ten-year intervals since 1790. Make a graph.

Materials Needed for LIFEPAC

Required:

Encyclopedia or Chemical
  ReferenceBook
metric ruler
small block of wood
string
small rock
water
graduated cylinder
balance
4 jars
4 iron nails
vinegar
ammonia
lemon
water
labels
sugar
test tube
fire source (Bunsen burner)
test-tube holder
salt
sand
measuring cup
filter paper

Suggested:

rock or mineral
Celsius thermometer
3 beakers (250 ml)
hot plate
3 cups of sugar
spoon
cotton thread
paper clip or tack
pencil
2-250 ml beakers
marbles
BB's or gravel about $\frac{1}{8}$-inch
  diameter
8th Grade Science Experiments video

Additional Learning Activities

Section I Properties of Matter

1. Demonstrate various crystal shapes using a good magnifying glass or a microscope. Common crystals are salt, boric acid, epsom salts, and sugar.
2. Using sugar on a plate (or in a beaker) add a small amount of sulfuric acid (H_2SO_4). Observe the oxidation of sugar without fire.

DO NOT TOUCH THE RESULTING CARBON UNTIL IT IS COOL AND RINSED.

3. Weigh out 5 grams of salt. Record the weight and volume. Repeat with 10 grams of salt. Repeat with 15 grams of salt. Make a line graph. What pattern do you see?
4. Water boils at 100°C at sea level. At what temperature does water boil where you are? Do not use a thermometer to stir.
5. Collect samples of ten different liquids and classify them according to their special properties.
6. Freeze 10 ml of water. What is its volume frozen? What happens when a bottle of soda is frozen?

Section II Atoms and Molecules

1. Draw an atom naming the parts. Include the orbitals.
2. Draw a water molecule. One is shown in your LIFEPAC. Be certain to measure the angle made by the two hydrogen atoms with the nucleus.

Section III Elements, Compounds, and Mixtures

1. Study the appearance of iron filings and powdered sulfur. Mix them together. Can you separate them? Try water and filter paper. Try a magnet. Which worked?
2. Heat the iron filings and sulfur. Can you separate them with a magnet now?
3. Make a poison poster illustrating common household chemicals that are poisonous and tell what to do if they are swallowed. Have other students check their homes and add to the list.
4. Check the Periodic Table to see if the mass of each atom is in the same order as the atomic number. If some are out of order, identify them.
5. Using information from the Periodic Table draw and label atoms of carbon, oxygen, sodium, chlorine, aluminum, and neon.

Materials Needed for LIFEPAC

Required:
paper, lump of sugar
soda cracker
sugar, water
dish, teaspoon
2 beakers, ice, heat source
cloth or styrofoam cup
water, balance
time piece with second hand
plaster of Paris, 2 stirrers
2 glasses or cups, 2 thermometers
flour, small candle, matches
flat candle base or watch glass
red cabbage, knife, pot, hot plate
small jar, crayons, vinegar
ammonia, lemon juice, milk, spoon
baking soda, 6 small test tubes
blue litmus paper, orange juice
cider, sugar water
salt water, liquid soap
baking soda in water
phenolphthalein
eyedropper, milk of magnesia
soapy water
red litmus paper
2 small paint brushes
soapy water or ammonia

Suggested:
asbestos pad
ring stand
8th Grade Science Experiments video

Additional Learning Activities

Section I Matter and Change

1. Using two different liquids (example: water and syrup), demonstrate the different boiling points of various liquids.
2. With a friend make a list of as many physical and chemical changes as you can think of.
3. In a chemistry book from the library, find an experiment that you can perform in class. Ask your teacher for the materials you need.

Section II Acids

1. Have students bring in products from home and test them with litmus paper.
2. Visit a chemistry lab. Make a one-page report of your observations.

Section III Bases

1. Have a discussion of the many uses for the different bases.
2. With your friends make a list of as many bases as you can think of. Use science books from the library.

3. Using science books from the library, write a one-page report on the difference between acids and bases.

Section IV Salts

1. Prepare a cup of hot tea and have the students describe as many physical properties as they can. Add a teaspoonful of lemon juice to the tea. Have the students describe any changes in the tea. Now add a teaspoonful of baking soda. Have the students describe the additional changes.

2. Go to the store with your friends. Find as many products on the shelf that refer to pH as you can. Write them on a piece of paper. Compare lists with other groups.

3. Write a one-page report on the use of salts. Use reference books from the library to help you.

Materials Needed for LIFEPAC

Required:
heat source
cooked egg white
bits of cheese
dried milk
egg yolk, any fruit
bits of green or yellow vegetables
soda cracker
white bread
paper towel
Tes Tape (from drugstore to test
   for sugar diabetes)
aluminum foil, potato slice
iodine (with dropper)
bits of various foods
color sections of newspapers or
   magazines,
file cards

Suggested:
marking pens
banner paper
encyclopedia
tape recorder and tape

Additional Learning Activities

Section I Foods and Digestion

1. Draw a diagram of the digestive system. Have your friends label the diagram.
2. Write a one-page report on vitamins. You may use library reference materials.

Section II Diet

1. Have the students make a bulletin board on MyPlate. The students can use the file cards they made.
2. Write to the United Dairy Council and ask them to send you nutritional information about MyPlate or download the information that is available online. Share and discuss the information with your friends.
3. Plan a dinner menu for your family using MyPlate.

Section III Nutritional Diseases

1. Discuss with the class the importance of eating natural foods. Ask them to name and write twenty-five natural foods. Discuss the differences in the lists.
2. Go to a natural foods store with a friend. Compare the differences between a regular grocery store and a natural foods store.
3. Write a one-page report on food additives. Use the library for reference materials.
4. Write a list of nutritional diseases and their causes. Check your diet and make sure you are getting the right nutrients.

Section IV Hygiene

1. Discuss the importance of sanitation in disease prevention. Invite an employee of the Health Department to speak to the class on the importance of personal and community hygiene.
2. Visit a local hospital with some of your friends. Report back to the class your impressions of the hospital hygiene.

Materials Needed for LIFEPAC

Required: Suggested:
none   several current science magazines

Additional Learning Activities

Section I Mechanical Energy

1. Have the students demonstrate several examples of kinetic and potential energy to the rest of the class.
2. Gather several pictures and tell whether the energy used in the picture is kinetic or potential.
3. Look up *mechanical energy* in the library. Make a list of all the professions you can think of that depend upon mechanical energy.

Section II Other Forms of Energy

1. Invite a local firefighter to speak to your class on spontaneous combustion and how it can be prevented.
2. Have students bring in recent articles on nuclear power. Discuss the advantages and disadvantages of nuclear power.
3. With your classmates make a bulletin board showing several examples of heat, chemical, and atomic energy.
4. With a friend visit the local electric company. Make a report on what you learned.
5. Ask your parents how the heat energy in your house is produced.

Section III Energy Conversion and Entropy

1. Lead a class discussion on the implications entropy has for a Christian.
2. Take a survey in your class to see how each student's house is heated or cooled. Make a chart to show how many students use electricity, how many use gas, and how many use solar heating and cooling.
3. Make a list of fifty ways to conserve energy. Call your electric company for ideas.

Materials Needed for LIFEPAC

Required:
world globe
flexible ruler
needle
glass
water
2 bar magnets
horseshoe magnet
iron fillings
sheet of clear glass or plastic
glass wand
silk
rubber wand or comb
fur
2 pith balls or 2 balloons

Suggested:
encyclopedias
hot dog
aluminum foil
cardboard
2 nails
scissors
glue
8th Grade Science Experiments video

Additional Learning Activities

Section I Magnetism

1. Make an electromagnet with a 1.5 volt battery, a piece of iron rod, and some bell wire. Measure the strength in terms of number of nails that can be picked up. Change the number of turns on the coil and measure again. Repeat as often as desired.
2. Make an electromagnet and use a compass to determine its north and south poles.
3. Make a coil of wire around a pencil or ruler. Make at least ten coils. With the long ends of this wire wrap five loops around a compass. Twist the ends together. Move a bar magnet in and out of the coil and record the activity of the compass needle.
4. Float a bar magnet on wood or cork. Place it in water. Diagram the way it points. Move the magnet on its float in the water. What happens?
5. Cut a number of cardboard pieces 2.5 cm square. Stack them up one by one with the N pole of a magnet on top. How many squares do you have before the magnet will not pick up a paper clip?

Section II Electricity

1. Connect a wire to a battery, then to a light bulb, and then to a material to be tested and back to the battery. Test a variety of materials and observe the light bulb. Does it light, shine dimly, or remain dark? Test wood, glass, chalk, paper clip, cardboard, copper, aluminum, or others as available.

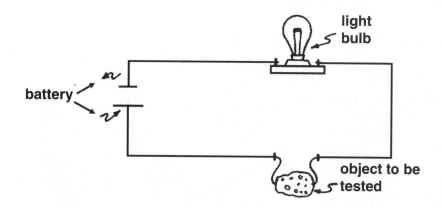

2. Use the same apparatus as above. This time try the open ends in liquids such as distilled water, tap water, salt water, vinegar, ammonia, alcohol, and other common liquids. Which are good conductors, poor conductors, and nonconductors?
3. With members of your family examine appliances and determine the number of watts each uses. Time the use of each for one day. Calculate the number of kilowatts.
4. Have someone explain to you how to read an electric meter and an electric bill.
5. Keep a record of miles driven in family cars and of the amount of gas purchased for two weeks. Find the number of miles per gallon of gas for each car.

Section III Energy for the Future
1. Visit a generating plant of any type.
2. Write to the Energy Research and Development Administration, for information to share with students about nuclear and solar power.
3. Illustrate "the domino theory" of a chain reaction by setting up dominoes spaced so all are toppled (see the following example).

X X X X X
X X X X
X X X
X X
X

4. Make and distribute posters urging safe and efficient use of energy.
5. Examine your present life style and determine how you would get along without electricity.

Materials Needed for LIFEPAC

Required:
strips of paper about 5 cm wide and
  at least 62 cm long
thin cardboards ($8\frac{1}{2}$ " by 11")
rubber bands, paper clips
25 identical washers
watch or stopwatch
tape measure or yardstick
scale to weigh students
scissors, tape, string

Suggested:
encyclopedias
calibrated force scale
cloth
hard cover book
aluminum foil
cardboard

Additional Learning Activities

Section I Distance
1. With a friend, measure the width of your classroom, first in meters and then in feet and inches. Ask another friend to make the conversion from meters to feet and inches.
2. With your classmates make a scale drawing of your classroom. When completed, color and decorate it and put it on the wall.
3. Write a one-page report on the importance of using the metric system in the United States.
4. Draw to scale a tree in your yard or neighborhood by measuring the angle that the line of sight to the top of the tree makes with the ground.

Section II Force
1. Make a list on the board of examples for each of Newton's laws of motion. Have the students complete the list.
2. Make up vectors for your friends. Have your friends calculate the sum of the vectors.
3. Research the life of Newton in the library. Write a one-page report on his scientific discoveries.

Section III Work
1. Using a small cardboard box filled with two pounds of sand, demonstrate the measuring of work with a spring balance.
2. Work several problems on the board using the formulas
$$w = F \times D \quad \text{and} \quad power = \frac{force \times distance}{time} = \frac{work}{time}$$
3. Many measurements in science are named after famous scientists (example: James Watt--watt). With a friend make a list of as many measurements named after famous scientists as you can think of.
4. Show how moving water is put to work. With the class make a water wheel using two pieces of thin wood four inches wide and ten inches long for the paddles. Fit the paddles into a U-shaped frame support. Hold the paddles in place with two thin nails. Let water fall on the paddles.
5. Using a reference book from the library, write a one- to two-page report on James Watt.
6. Use a reference book to learn how dams and waterfalls help humans harness energy. Give a report to the class.

Materials Needed for LIFEPAC

Required:
coins (pennies are best)
8 or 10 round sticks
 (pencils, Tinker Toys, dowels)
2 blocks of wood
bar of soap
pencil sharpener (wall-mounted)
$2\frac{1}{2}$ meters of string
sturdy coat hanger
spool
wire cutters
rubber band (medium size)
matchbox full of pebbles
2 rulers of different lengths
3 or 4 books

Suggested:
automotive magazines
tape
eggbeater

Additional Learning Activities

Section I Friction

1. Demonstrate several examples of friction (examples: striking a match, pushing a chair across the floor).
2. Lead a class discussion on the advantages and disadvantages of friction.
3. See how friction is reduced. Slide two pieces of rough wood over each other. Then sand both pieces smooth and slide again. Is the friction less?
4. Slide two flat pieces of metal together. Then put a little grease or oil between them. Is the friction less?

Section II Levers

1. Demonstrate the use of a lever to your class.
2. With a classmate prepare an exhibit of small machines. Label each machine and tell how it works.
3. Make a list of as many common levers as you can think of. Make a drawing of each type of lever.

Section III Wheel and Axle, Pulleys, and Gears

1. Demonstrate the use of a wheel and axle, a pulley, and gears to your classroom.
2. Ask the class to list the different uses for axles, pulleys, and gears. Compile the lists on the board. Ask the class if they can think of any further uses.
3. List all the machines that you use during one week. Turn in your list to your teacher.

Section IV Inclined Plane, Wedge, and Screw

1. Have the class collect pictures of different machines for a bulletin board. Label all machines.
2. Demonstrate the use of the inclined plane, wedge, and screw.

Materials Needed for LIFEPAC

Required:
microscope
single-edged razor blade
glass slides and cover slips
leaves from two unlike plants
250-milliliter beaker of freshly
  picked leaves
fastener
large glass jar for terrarium
assorted plants
stones, soil, insects
popped popcorn
plastic bags
three pans
grass seed
watering can
ruler, water

Suggested:
encyclopedias

Additional Learning Activities

Section I Photosynthesis and Food
1. Demonstrate the use of the microscope as well as the care and cleaning of the parts. Allow students to look at a variety of items of their own choosing until their original amazement is satisfied.
2. Plan a noon lunch of vegetable and fruit salads. Use the chart from Section I and discuss the plant parts served.
3. Grow a house plant from cuttings

Section II Natural Cycles
1. Discuss the problems of water in your area: pollution, flooding, drought, irrigation, and so on.
2. Make a compost pile of grass clippings and leaves from the lawn of your school or church.
3. Make an attractive terrarium from a recycled jar and give it as a gift. Try not to purchase anything for the terrarium. It should be all your efforts.
4. Gather five organic materials such as a leaf, orange peel, paper towel, or potato slice. Place each in a plastic bag with a few drops of water. Seal tightly and observe the decay process for one week. Keep good records with drawings.

Section III Balance and Disruption
1. Discuss ecology problems of your area.
2. List the natural resources of your area or state. Discuss the conservation of each.
3. Visit a local habitat (even a vacant lot or park will work) and make a list of every animal, plant, and insect. Try to determine what each organism eats and what eats it.
4. Sketch a food chain for three carnivores of your area.

Materials Needed for LIFEPAC

Required:
unidentified items in a box
beaker or cup
dish to catch water
graduated cylinder
magnet
bottle containing a material prepared
  by the teacher
toothpicks
a box of dominoes
bar magnet
paper
wood
iron filings
other unmagnetized materials
string
block of wood
board 2-3 ft. long
piece of metal or similar objects of different weights
spring balance

Suggested:
dish
pressure cooker
heat source
ringstand
screen
thermometer
ammonia
jar

Additional Learning Activities

Section I Basic Science

1. Discuss the differences between observation, investigation, and experimentation. Review examples of each.
2. Discuss how the structure of the metric system makes it different from the standard system.
3. Have students define an element and describe its physical properties and how it compares to other elements. Pupils may choose an element to investigate. They should keep track of how they went about finding information. A collection of elements would make a good future classroom resource.
4. Use a world map or U.S. map and locate where ten to twenty-five elements are found. Label locations on the map and prepare for classroom display.

Section II Physical Science

1. In what ways is heat produced by the conversion of energy? Discuss and demonstrate heat produced by mechanical energy (friction), by chemical energy, solar energy, atomic energy, and electrical energy (light bulb burning).
2. Visit a manufacturing plant to see machines in operation.
3. Use merchandise catalogs that may be cut up as a source of pictures to illustrate each of the simple machines in use individually or as part of complex machines. Make a poster and label the simple machines.
4. Collect toys that use simple machines and make a display. Be able to explain how they are simple machines.

5. Individually or with a partner make one or more of the following items: a fuse, a switch, a telegraph set, or an electric motor. Use easily obtained materials.

6. Some pupils are mechanically adept and would enjoy constructing and explaining a fairly complicated electrical or mechanical display. Discuss plans with pupils, before permitting construction of the special project.

Section III Life Science

1. How is the body a machine? Discuss food as fuel in the body, bones as levers, and work accomplished by structural parts.

2. How do drugs, alcohol, and smoking affect the body? Would this be "defiling the temple of God"?

3. Africa is a developing continent. What is happening to animal and plant life as people claim more of the land? Is the balance of nature being upset?

4. Visit a museum or display that features the structure of the human body and health principles.

5. Student(s) may be interested in vocations in the health field. Have them investigate possibilities and prepare a poster for display. Arranging to visit a hospital to observe personnel at work would be a valuable experience.

Section IV Vocations in Science and Technology

1. As a pupil, how can you look to the years ahead and draw up a plan to develop yourself as a person and a good worker:

2. How do you fill out an application? Teacher can supply each pupil with an application obtained from a local business and pupils practice filling it out.

3. Why is a good character essential to getting the right job?

4. How should an applicant act during an interview?

5. A student(s) should be assigned the task of reviewing newspaper calendars; local college and school calendars; and any other sources for talks, films, slides, or other aids dealing with vocations and careers. The pupil doing the research should be given the responsibility to notify the class when there are opportunities to attend.

SCIENCE RECORD

Observation # _____   Name _____
Investigation # _____   Date _____
Experiment # _____

PROBLEM:_____
_____
_____
_____

MATERIALS:_____
_____
_____

METHOD:_____
_____
_____
_____
_____
_____
_____
_____
_____
_____
_____

RESULT:_____
_____
_____
_____
_____

CONCLUSION:_____
_____
_____
_____
_____
_____

Notes

ALTERNATE

TESTS

Reproducible Tests
for use with the Science 800
Teacher's Guide

Name _____

Answer *true* or *false* (each answer, 1 point).

1. _____ Aristotle was an ancient philosopher.
2. _____ Newton used mathematics to develop the Law of Universal Gravitation.
3. _____ The theory of LaMarck was disproved.
4. _____ The number $6.23 \cdot 10^4$ is the same as 62.34.
5. _____ Experiments are used to prove a hypothesis.
6. _____ Polio vaccine was developed by Salk.
7. _____ Copernicus invented the wheel.
8. _____ The first book printed on Gutenberg's press was the Bible.
9. _____ The production of cotton increased when the light bulb was invented.
10. _____ Our imperfect technology is a cause of pollution.

Solve these problems (each answer, 3 points).

11. Write $5.34 \cdot 10^2$ in numerals. _____
12. How many millimeters make one meter? _____
13. Add and write the answer with the proper number of significant figures.

$$\begin{array}{r} 8.3 \\ 4.56 \\ +6.83 \\ \hline \end{array}$$

Match these terms (each answer, 3 points).

14. _____ technology
15. _____ biodegradable
16. _____ Bible
17. _____ radium
18. _____ theory
19. _____ $3.124 \cdot 10^2$
20. _____ cancer cure

a. scientifically correct
b. more certain than a hypothesis
c. a goal of life science
d. scientific notation
e. Marie Curie
f. capable of being broken down by the action of bacteria
g. metric system
h. applied science

Science 801 Alternate Test

Complete these statements choosing from the terms listed below (each answer, 3 points).

| shaduf | Copernicus | solar energy |
| God | coal | Einstein |

21.   One in control of everything is _____.
22.   The Egyptians developed the _____
      for irrigation.
23.   A non-polluting, safe form of energy is _____ .
24.   _____ developed the equation, $E = mc^2$.
25.   Galileo agreed with the theory of _____
      that the earth was not the center of the universe.

Answer these questions (each answer, 3 points).
26.   What are three problems of modern society?
      a. _____
      b. _____
      c. _____
27.   What was the result of Johann Gutenberg's invention?
      _____
      _____
28.   What are three benefits of modern technology?
      a _____
      b. _____
      c. _____

61/76

Date _____
Score _____

Name _____

Match these items (each answer, 2 points).

1. _____ atom   a.   amount of matter in an
2. _____ mixture   object
3. _____ compound   b.   burning
4. _____ physical property   c.   basic unit of matter
5. _____ chemical property   d.   early theory of atom
6. _____ volume   e.   mass ÷ volume
7. _____ density   f.   color
8. _____ mass   g.   gravitational pull
9. _____ Dalton   h.   atom particle in the nucleus
10. _____ proton   which is positively charged
                                         i.   a union of elements
                                         j.   a combination of several
                                              kinds of materials
                                         k.   l x w x h

Answer *true* or *false* (each answer, 1 point).

11. _____ Elements are forms of matter with two or more types of atoms.
12. _____ All liquids are composed of matter.
13. _____ Water boils at 100° on the Celsius scale.
14. _____ Electrons have no electrical charge.
15. _____ Protons orbit around the nucleus.
16. _____ All matter is made up of basic units called atoms.
17. _____ Gas pressure increases when gas molecules are heated.
18. _____ Density is a chemical quality.
19. _____ Compounds can be separated into elements by dissolving them
             in water.
20. _____ The Periodic Table is an alphabetical listing of the elements.

Write the correct answer on each line (each answer, 3 points).

21.   An example of an element is _____.
      a. air   b. sodium   c. table salt
22.   Atomic mass is a measure of _____.
      a. protons and neutrons   b. charge   c. electrons
23.   A form of matter with no definite volume and no definite shape
      is _____.
      a. solid   b. liquid   c. gas
24.   In orbits around the nucleus are found _____.
      a. protons   b. neutrons   c. electrons
25.   Scientists have discovered more than _____ different
      elements.
      a. 30   b. 50   c. 100

Put an X before each item that is matter and an O before each item that is *not* matter (each answer, 1 point).

26. _____ earth   31. _____ wheat
27. _____ heat   32. _____ electricity
28. _____ air   33. _____ magnetism
29. _____ water   34. _____ steak
30. _____ energy   35. _____ apple

Answer these questions on the basis of the Periodic Table entry on carbon (each answer, 2 points).
36. What is the atomic mass? _____
37. What is the atomic number? _____
38. How many neutrons does carbon have? _____
39. What is the chemical symbol of carbon? _____
40. How many protons does carbon have? _____
41. How many electrons does carbon have? _____

Count the total number of atoms in each formula (each answer, 3 points).

42. H_2O _____   45. Na_2CO_3 _____
43. CO_2 _____   46. H_2SO_4 _____
44. NaCl _____

66/82

Date _____
Score _____

Name _____

Define the following words (each answer, 5 points).
1.   evaporation _____
2.   condensation _____
3.   distillation _____

Match the following items (each answer, 2 points).
4.   _____ oxidation
5.   _____ phenolphthalein
6.   _____ base
7.   _____ fusion
8.   _____ slow oxidation
9.   _____ hard water
10. _____ H$^+$
11. _____ fission
12. _____ Geiger counter
13. _____ strong acid

a.   acid
b.   splitting of the nucleus
c.   instrument used to detect radioactivity
d.   uniting of two nuclei
e.   pH of 1-2
f.   neither acid nor base
g.   high in mineral salts
h.   an indicator
i.   a substance combining with O$_2$
j.   rusting
k.   OH$^-$

Answer *true* or *false* (each answer, 1 point).
14. _____ Adding heat to a substance increases its mass.
15. _____ Burning is an example of a chemical change.
16. _____ Acids feel slippery and taste bitter.
17. _____ Hydronium is water plus a hydrogen.
18. _____ Sodium hydroxide is an example of a base

Identify acids and bases (each answer, 2 points).
19. Write *B* after each base. Write *A* after each acid. Write *N* after each item that is neutral.

a.   HCl   _____
b.   NaOH   _____
c.   Ca(OH)$_2$ _____
d.   pure water   _____

e.   orange juice   _____
f.   milk of magnesia   _____
g.   ammonium hydroxide _____
h.   sulfuric   _____

Complete these lists (each answer, 3 points).

20. List three uses of salts.

a. _____

b. _____

c. _____

21. List three uses of acids.

a. _____

b. _____

c. _____

59	
	74

Date _____

Score _____

Name _____

Answer *true* or *false* (each answer, 1 point).

1. _____ Carbohydrates contain carbon, hydrogen, and sugar.
2. _____ Amino acids are found in fats.
3. _____ Spiritual health is as important as physical health.
4. _____ Protein aids in production of enzymes.
5. _____ Iodine helps control the rate of absorption.
6. _____ The first step of digestion takes place in the mouth.
7. _____ Niacin is Vitamin B_3.
8. _____ Rickets are caused by a Vitamin C deficiency.
9. _____ No relationship exists between allergy and addiction.
10. _____ Allergies have been studied for years by scientists.

Complete these lists (each answer, 3 points).

11. List the four major minerals necessary in a daily diet.

a. _____ c. _____
b. _____ d. _____

12. List the five food groups of MyPlate.

a. _____ d. _____
b. _____ e. _____
c. _____

13. List the six basic nutrients necessary for a balanced diet.

a. _____ d. _____
b. _____ e. _____
c. _____ f. _____

Match these items (each answer, 2 points).

14. _____ nutrient containing amino acids
15. _____ mineral that helps control metabolism rate
16. _____ helps prevent anemia
17. _____ prevents scurvy
18. _____ helps clot blood
19. _____ helps cells function normally
20. _____ carries oxygen in red blood cells
21. _____ mineral needed for bones and teeth
22. _____ needed for healthy skin and eyes
23. _____ oily, greasy animal or vegetable substance
24. _____ foods containing carbon, oxygen, and hydrogen
25. _____ liquid part of the blood
26. _____ organic substances necessary for normal growth and health
27. _____ necessary substances, neither animal nor vegetable, in diet

a. vitamins
b. carbohydrates
c. Vitamin K
d. fats
e. Vitamin E
f. protein
g. Vitamin D
h. minerals
i. Vitamin C
j. B vitamins
k. phosphorus
l. Vitamin A
m. iodine
n. iron
o. plasma

Complete the following statements (each answer, 3 points).

28. The digestion of starch begins in the _____.
29. The breakdown of protein begins in the _____.
30. Nutrients are absorbed into the bloodstream in the _____.
31. A substance that hinders bacterial growth is called a(n) _____.
32. The cause for much disease in the past was the lack of_____.

Date _____

Score _____

Name _____

Answer *true* or *false* (each answer, 1 point).
1. _____ Force is a push or a pull.
2. _____ A pendulum at the top of its arc is an example of kinetic energy.
3. _____ Heat loss can be prevented by insulation.
4 _____ Most atomic particles are found in the nucleus of the atom.
5 _____ Neutrons have a positive charge.
6. _____ Heat is a form of kinetic energy.
7. _____ Electricity is created by generators.
8. _____ Friction produces heat.
9. _____ Matter is anything that occupies space.
10. _____ Radiant energy is energy of motion.

Write the letter for the answer on the line (each answer, 2 points).
11. Heat is a form of _____ energy.
   a. potential   c. nuclear
   b. kinetic   d. atomic
12. The rubbing of molecules against one another is _____.
   a. work   c. conduction
   b. atoms   d. friction
13. The measurement of heat is called _____.
   a. temperature   c. conduction
   b. convection   d. current
14. The particle with a positive charge (+) is called a _____.
   a. neutron   c. electron
   b. proton   d. atom
15. The energy of electrons is called _____ energy.
   a. electrical   c. nuclear
   b. thermal   d. atomic
16. The measure of order and disorder in the world is called _____.
   a. energy   c. thermo
   b. randomness   d. entropy

Match these items (each answer, 2 points).

17. _____ God
18. _____ work
19. _____ kinetic energy
20. _____ potential energy
21. _____ radiation
22. _____ insulator
23. _____ fission
24. _____ thermo
25. _____ exothermic
26. _____ generator

a. heat that travels in waves
b. reaction in which heat is given off
c. splitting
d. force moving through a distance
e. source of the universe's energy
f. heat
g. randomness
h. prevents heat loss
i. stored energy
j. creates electricity
k. moving energy

Complete these items (each answer, 5 points).

27. Explain the difference between kinetic and potential energy.

_____
_____
_____

28. Explain the difference between force and work.

_____
_____
_____

Date _____

Score _____

Name _____

Match these items (each answer, 2 points).

1. _____ semiconductors
2. _____ insulators
3. _____ magnetometer
4. _____ coal
5. _____ volt
6. _____ ohm
7. _____ lodestone
8. _____ conductor
9. _____ circuit
10. _____ generator
11. _____ geothermal
12. _____ watt

a. material that prevents the flow of electricity
b. unit of electrical resistance
c. abundant fuel
d. unit of electrical power
e. unit of electrical potential
f. instrument used by geologists searching for oil
g. crystals
h. heat from the earth
i. material that permits the flow of electrons
j. unit of gravity
k. path of moving electrons
l. changes mechanical energy to electrical energy
m. a rock or mineral that is naturally magnetic

Write the correct letter for the answer on the line (each answer, 2 points).

13. Objects with unlike charges _____ .
 a. repel each other   c. attract each other
 b. are neutral to each other   d. none of these

14. Charged objects _____ .
 a. attract neutral objects   c. attract each other
 b. repel each other   d. none of these

15. The man who built the first A.C. generating plant in the United States was _____ .
 a. William Gilbert   c. George Westinghouse
 b. Thomas Edison   d. Benjamin Franklin

16. The man who proved that lightning is a form of static electricity was _____ .
 a. Benjamin Franklin   c. Thomas Edison
 b. Michael Faraday   d. Alessandro Volta

17. The man who built the first working electric light bulb was _____ .
 a. Georg Ohm   c. Thomas Edison
 b. Hans Christian Oersted   d. Michael Faraday

18. The man who built a battery was _____ .
 a. Alessandro Volta   c. Andre-Marie Ampere
 b. Georg Ohm   d. Charles Coulomb

19. A high resistance copper wire for heating is _____ .
 a. copper   c. carbon
 b. nichrome   d. zinc

20. The rate of energy use is measured in _____ .
    a. amperes   c. ohms
    b. volts   d. watts

21. The splitting of heavy atoms into two or more light atoms is called

    _____ .
    a. fission   c. transmission
    b. fusion   d. reactor

22. The author of *De Magnete* is _____ .
    a. Oersted   c. William Gilbert
    b. King Oscar   d. James Watt

23. The type of circuit shown in the illustration _____ .

    a. open   c. parallel
    b. closed   d. series

24. The type of circuit shown in the illustration is _____ .

    a. parallel   c. open
    b. series   d. closed

25. "The strength of a field decreases as the distance increases." This is a
    statement from _____ .
    a. Ohm's Law   c. Einstein's Theory
    b. Westinghouse Hypothesis   d. Inverse Square Law

Answer *true* or *false* (each answer, 1 point).

26. _____ The unit of current is the ampere.
27. _____ Solar energy produces most of the energy in the United States today.
28. _____ Falling water is used to produce electricity.
29. _____ Semiconductors make possible miniature radios.
30. _____ A kilowatt is equal to 1,000 volts.
31. _____ The United States has all the gas it needs.
32. _____ Infrared is invisible energy used to make light.
33. _____ Putting two light atoms together to make a single atom is fusion.
34. _____ Temporary magnets can be made using electricity.
35. _____ When electrons are removed from an object it is said to be charged positively.
36. _____ Solar furnaces are low heat toys.
37. _____ Iron filings can be used to locate magnetic lines of force.
38. _____ Ohm's Law is "volts equal ohms times amperes."
39. _____ Benjamin Franklin made an arc light.
40. _____ When protons are taken away, an object has a negative charge.

Complete these activities (each answer, 5 points).

41. Calculate the electromotive force required to push 5 amperes through a 22 ohm resistance.

42. Calculate the power supplied in a circuit with a current of 7 amperes with 110-volt service.

Date _____

Score _____

Name _____

Match these items (each answer, 2 points).

1.	_____	ancient measure of length	a.	force
2.	_____	work divided by time	b.	inertia
3.	_____	push or pull	c.	cubit
4.	_____	ability to do work	d.	energy
5.	_____	a form of mathematics	e.	meter
6.	_____	measure of length	f.	geometry
7.	_____	tendency of mass to resist change	g.	work
			h.	power

Answer *true* or *false* (each answer, 1 point).

8. _____ The English system uses meters for measuring length.
9. _____ Gravity is the force of the earth on objects near it.
10. _____ Velocity is an example of a vector quantity.
11. _____ Galileo believed in God.
12. _____ Work depends on force and power.
13. _____ You cannot see energy.

Complete these statements (each answer, 3 points).

14. A rough calculation is called a(n) _____ .
15. The metric system was devised by _____ .
16. Vector quantities have both magnitude and _____ .
17. The metric unit of force is the _____ .
18. Work is measured in _____ .
19. Kinetic energy is the energy of _____ .
20. The three laws of motion were discovered by _____ .
21. A force has direction and _____ .
22. If a quantity has magnitude only, it is a _____ quantity.

Complete these items (each answer, 5 points).

23. Use the scale drawing to find the actual distance of AB.

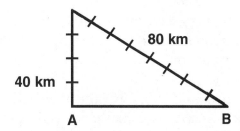

24. Construct the vector sum of:

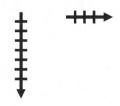

25. Calculate the power developed by a motor that moves an object 10 m against a force of 20 N in 5 seconds.

Date _____

Score _____

Name _____

Answer *true* or *false* (each answer, 1 point).
1. _____ Friction is a force.
2. _____ Liquids and gases are examples of fluid friction.
3. _____ The human forearm is an example of a third class lever.
4. _____ The product of a force and its distance to the fulcrum is called torque.
5. _____ The wheel and axle have the same axis.
6. _____ Gears the same size change direction and speed.
7. _____ The efficiency of a good machine is always 100 percent.
8. _____ Jesus' carpenter tools were made from simple machines.
9. _____ The inclined plane is used to raise heavy objects.
10. _____ Wedges have been used throughout history.

Write the correct letter of the answer (each answer, 2 points).
11. Any substance that flows is called _____ .
    a. gas   c. fluid
    b. liquid   d. water
12. The support on which a lever turns is the _____ .
    a. fulcrum   c. torque
    b. wheel   d. machine
13. An example of an inclined plane is a _____ .
    a. pulley   c. wheel
    b. ramp   d. baseball bat
14. Machines are made less efficient by _____ .
    a. AMA   c. torque
    b. IMA   d. friction
15. The force that opposes the start of motion is referred to as _____ friction.
    a. starting   c. rolling
    b. skidding   d. normal

Complete these items (each answer, 5 points).

16. A lever is 15 meters long. A load is placed at one end 3 meters from the fulcrum. The load is 3,000 newtons. The effort force of 2,000 newtons is 6 meters from the fulcrum.

   a. Calculate the IMA

   b. Calculate the AMA

   c. Calculate the efficiency in percent.

17. State which wedge will have a higher IMA. _____

 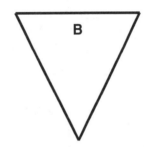

18. A box weighing 24 newtons requires a force of 6 newtons to push it along the floor. Calculate the coefficient of friction.

36/45

Date _____

Score _____

Science 809 Alternate Test

Name _____

Match these items (each answer, 2 points).
1. _____ hybrid
2. _____ chlorophyll
3. _____ ground water
4. _____ nitrogen
5. _____ oxygen
6. _____ guard cell
7. _____ natural control
8. _____ environment
9. _____ epidermis
10. _____ herbivore

a. controls opening of a stomata
b. conditions that limit population numbers
c. eats only plants
d. waxy outer cells of plants
e. the surroundings of an organism
f. bacteria and fungi
g. the result of a cross between two unlike pants or animals
h. given off in photosynthesis
i. green pigment
j. stored beneath surface of earth
k. necessary for green foliage and plant growth

Answer *true* or *false* (each answer, 1 point).
11. _____ Greed causes humans to kill beautiful birds and animals.
12. _____ Trace elements are needed in great amounts.
13. _____ Protein is necessary for good health.
14. _____ Transpiration is loss of water through stoma.
15. _____ Humans are often omnivores.
16. _____ Hybrid plants often have greater yields.
17. _____ Many underdeveloped nations are overpopulated.
18. _____ Decay of material is always bad.
19. _____ Legumes are a type of fungi.
20. _____ Constant use of chemical sprays has caused the development of some resistant strains of insects.

Write the correct for the answer on the line (each answer, 2 points).

21. Legumes are very important because of _____ .
    a. rhizobium bacteria   c. hybrids
    b. precipitation   d. disease resistance

22. Bacteria and fungi are important agents of _____ .
    a. hybridization   c. photosynthesis
    b. chlorophyll production   d. decay

23. A fertilizer label reads 30-19-11. This means _____ .
    a. 30% water, 19% air, 11% soil
    b. 30% CO_2, 19% O_2, 11% N_2
    c. 30% nitrogen, 19% phosphorus, 11% potassium
    d. 30% chlorophyll, 19% water, 11% carbon

24. The three major advantages of modern agriculture are machinery, agricultural chemicals, and _____ .
    a. McCormick's reaper   c. hybrid plants
    b. chloroplasts   d. phosphorus

25. Photosynthesis requires sun, chlorophyll, and _____ .
    a. glucose   c. protein
    b. sucrose   d carbon dioxide

26. In times of food shortage it is best to _____ .
    a. grow grain and feed it directly to humans
    b. feed grain directly to cattle
    c. feed grain directly to chickens
    d. grow corn and feed it directly to hogs

27. Natural resources include wildlife, wilderness, air, water, minerals, fossil fuels, _____ .
    a. soil, and forests   c. shelterbelts, and forests
    b. aluminum, and gold   d. coal, and oil

28. Human populations no longer are controlled by famine and _____ .
    a. food pyramids   c. industrial advances
    b. herbivores   d. disease

29. What do plants provide for animals? _____
    a. food and chlorophyll   c. oxygen and phosphorus
    b. food and oxygen   d. nitrogen and glucose

30. Development of solar energy would help conserve _____ .
    a. fossils fuels   c. top soil
    b. aluminum   d. endangered species

Match these items (each answer, 2 points).

31. _____ carbon dioxide
32. _____ recycling
33. _____ photosynthesis
34. _____ hybrid
35. _____ ecology
36. _____ agricultural chemicals
37. _____ decomposers
38. _____ nodule
39. _____ scavenger
40. _____ precipitation

a. fertilizers, insecticides, herbicides
b. feeds on dead plants and animals
c. root structure containing rhizobium
d. rain, snow, hail
e. bacteria and fungi
f. often has greater yield
g. given off in animal respiration
h. stomata
i. $6CO_2 + 12H_2O \xrightarrow[\text{chlorophyll}]{\text{light}} C_6H_{12}O_{22} + 6O_2 + 6H_2O$
j. reuse of old materials when making new products
k. study of organisms in relationship to their environment

56 / 70

Date _____
Score _____

Name _____

Match these items (each answer, 2 points).

1.	_____ identified by your senses	a.	science
2.	_____ keeps molecules together	b.	proficient
3.	_____ midpoint of pH scale	c.	scientific method
4.	_____ cannot dissolve in liquid		
5.	_____ elements are identified by	d.	science (pure)
6.	_____ energy of position	e.	matter
7.	_____ orderly study of natural and physical things	f.	density
		g.	electrolysis
8.	_____ carbon dioxide	h.	radiation
9.	_____ number below the letter	i.	attraction
10.	_____ knowledge, skills, or expertness	j.	symbols
11.	_____ taste of bases	k.	subscript
12.	_____ size of an object compared to its weight	l.	mixture
		m.	insoluble
13.	_____ sun rays reach earth by this method	n.	sour
14.	_____ nine steps for developing scientific principles	o.	acids
		p.	litmus paper
15.	_____ working to formulate scientific principles	q.	bitter
		r.	neutral
16.	_____ sand, salt, and sawdust	s.	adaptability
17.	_____ electric current passed through water	t.	CO_2
18.	_____ changing easily	u.	potential
19.	_____ taste of acids	v.	sweet
20.	_____ hydrogen is a common atom to		
21.	_____ how are acids and bases tested		

Match the metric unit to the standard unit given (each answer, 2 points).

22.	_____ 88 km/h	a. centimeter
23.	_____ yard of ribbon	b. kilogram
24.	_____ ton of coal	c. 55 MPH
25.	_____ quart of orange juice	d. meter
		e. liter

Answer *true* or *false* (each answer, 1 point).

26. _____ Packaged grocery items are marked by standard and metric measures.
27. _____ A cork floats on water because of its buoyancy.
28. _____ There is no space in the atom.
29. _____ All bases contain an OH ion.
30. _____ The lowest temperature reached by humans is -500 degrees Celsius.
31. _____ Winds are caused by convection currents.
32. _____ Generalizations are general statements used regularly by scientists.
33. _____ We are not responsible for other people on earth.
34. _____ Unlike electric charges attract each other.
35. _____ Static electricity flows through a wire.
36. _____ There are more protons in the atom than electrons.
37. _____ Current electricity is an object defined by its physical characteristics.
38. _____ The Bible teaches us to prepare for time of need.
39. _____ The Bible does not give evidence of simple machines.
40. _____ A machine needs proper fuel and so does the body.
41. _____ Vitamins produce heat to keep the body warm.
42. _____ The natural balance of nature can be destroyed by fire.

Complete these sentences (each answer, 3 points).

43. Heat is the energy of _____ molecules.
44. When ice forms on water it has a lower _____ and floats.
45. Neither matter nor energy can be a. _____ or b. _____ but they can be changed from one to the other.
46. Two sources of energy under development are a. _____ and b. _____ .
47. The apostle Paul said that if you did not want to work you should not _____ .

Answer these questions (each answer, 5 points).

48.	In what three ways do magnetism and electricity have something in common?

Magnetism	Electricity
a. _____	_____
_____	_____
b. _____	_____
_____	_____
c. _____	_____
_____	_____

49.	Why do strong smelling substances spread their odors through a room?

_____

_____

_____

50.	Describe what you think will happen if all controls on the use of the atomic bomb, hydrogen bomb, and nuclear warheads are lifted for all countries.

_____

_____

_____

_____

_____

_____

_____

_____

_____

_____

_____

_____

_____

91 / 113

Date _____
Score _____

Notes

LIFEPAC

ANSWER KEYS

71

SECTION ONE

1.1 Science is knowledge.
1.2 Science is orderly knowledge.
1.3 Science is orderly knowledge proved by experiments.
1.4 experiment
1.5 knowledge
1.6 Either order:
  a. true
  b. false
1.7 Orderly knowledge demonstrated by repeated experiments.
1.8 I would feed fish to several cats and if they ate them, my hypothesis would be proved.

1.9 b.

1.10 a.

1.11 c.

1.12 b.

1.13 false
1.14 true

1.15 false

1.16 They were not able to prove their ideas, so many of their ideas were false.

1.17 d

1.18 f

1.19 a
1.20 c
1.21 e

1.22 b
1.23 Since the Moors brought more advanced ideas, Western scientists would have been saved the time required to discover these ideas.
1.24 Renaissance
1.25 Copernicus
1.26 Galileo
1.27 Universal Gravitation

1.28 Sir Isaac Newton
1.29 true
1.30 false

1.31 false
1.32 true

1.33 false

1.34 true

1.35 true

1.36 teacher check

1.37 That pitchblende, an ore of radium, gives off radiation.

1.38 It means energy equals mass times the square of the speed of light. $E = mc^2$ is read, "energy equals mass times the square of the speed of light."
1.39 God is perfect and never makes the mistakes men do.

1.40 a. electron
  b. neutron
  c. proton
1.41 f
1.42 d
1.43 a

1.44	b	1.70	$4.3 \cdot 10^2$
1.45	Choose a problem.	1.71	$6.282 \cdot 10^3$
1.46	Make a hypothesis.	1.72	$5 \cdot 10^4$
1.47	Research what others have done.	1.73	$2.85 \cdot 10^2$
1.48	Perform experiments.	1.74	$7.96 \cdot 10^3$
1.49	If true, restate the hypothesis as a theory.	1.75	5,000
		1.76	3,230
1.50	If not true, state a new hypothesis and begin again.	1.77	582
		1.78	12,000
1.51	Write and publish a paper.	1.79	64,000,000
1.52	Change the theory should it be proved wrong.	1.80	4
1.53	Restate the theory as a law.		
1.54	c. A certain substance will kill a rat.	1.81	2
1.55	b. Similar substances have killed rats.	1.82	5
1.56	e. Give the substance to many rats.	1.83	2
1.57	a. The rats died.	1.84	1
1.58	d. State the theory of Rat-Kill.	1.85	7
1.59	g. Publish a paper.	1.86	630
1.60	f. State the law of Rat-Kill.	1.87	5,200
1.61	The use of the scientific method will help to ensure reliability of the findings and conclusions.	1.88	73
		1.89	8,500
		1.90	146.0
1.62	ten	1.91	1,007
1.63	defined	1.92	14.3569
1.64	derived	1.93	8
1.65	20	1.94	12.3
1.66	gram	1.95	2
1.67	one thousand	1.96	1
1.68	one-hundredth	1.97	4
1.69	one-thousandth	1.98	4

1.99   9
1.100 29
1.101-1.104   teacher check
1.105 approximately 1 kilogram, or
       1,000 grams

1.106 approximately 20 g

1.107 approximately 4,000 g

1.108 $8.2 \cdot 10^1$
1.109 $1.263 \cdot 10^3$

1.110 $1 \cdot 10^6$

1.111 $5.41 \cdot 10^2$

1.112 $2.000004 \cdot 10^6$

1.113 $1.063 \cdot 10^2$

1.114 $8.205 \cdot 10^2$

1.115 410
1.116 50,000,000,000
1.117 183,000
1.118 1,546.3
1.119 96,254.8

1.120 2
1.121 1
1.122 3
1.123 4

1.124 5

1.125 C

1.126 I
1.127 I

1.128 C

1.129 8.43

1.130 90,900

1.131 566

1.132 10,400

1.133 $4.8 \cdot 10^2$

1.134 $8.4 \cdot 10^6$

1.135 $5.5 \cdot 10^3$
1.136 4.3

SECTION TWO

2.1   true
2.2   false
2.3   true
2.4   true
2.5   false
2.6   false
2.7   false
2.8   false
2.9   true
2.10 true
2.11 c
2.12 b
2.13 c
2.14 a
2.15 c
2.16 e
2.17 a
2.18 b

2.19 Many ancient discoveries and knowledge were lost.

2.20 crossbow

2.21 wheel

2.22 water mill

2.23 Renaissance

2.24 God gave humans minds in order to fulfill the commandment to "subdue" the earth. We need to understand the Earth to subdue it.

2.25 b. printing press

2.26 a. the Bible

2.27 b. gunpowder

2.28 c. "animalcules"

2.29 b. Renaissance period

2.30 Answer should include some of the following ideas:
Drew plans for a flying machine and a parachute. Designed artillery and planned the diversion of rivers. Produced many drawings of machines and of experimental inventions.

2.31 Either order:
a. Thomas Newcomen
b. James Watt

2.32 The gin cleans by separating the seeds from the fibers.

2.33 Faraday discovered the principle of the dynamo, or generator.

2.34 Either order:
a. power stations
b. light bulb

2.35 Examples:
a. Science is supplying principles at a faster rate.
b. Communication of ideas is faster and easier.
c. People demand more conveniences.
d. Our society can support research as never before.

2.36 true

2.37 true

2.38 false

2.39 false

2.40 false

2.41 food would be more scarce than it is

2.42 of advances in medicine, improvement in food supply, and machinery that has been invented to aid life

2.43 a steadily increasing population

2.44 Any order:
a. food
b. warmth
c. clothing
d. shelter
e. air
f. amusement
g. communication
h. transportation

2.45 b. rapid transit

2.46 b. TV dinners

2.47 a. coal

2.48 c. cancer

2.49 a. industrial wastes

2.50 b. biodegradable

2.51 Technology is not to blame. Humans abuse technology. Technology is the source of solutions to community and environmental problems.

SECTION THREE

3.1	teacher check	3.11	d
3.2	false	3.12	b
3.3	true	3.13	a
3.4	false	3.14	c
3.5	true	3.15	Examples:
3.6	teacher check		To find a cure for cancer, to develop new food sources
		3.16	Examples:
3.7	are not perfect		To improve national defense, to explore space
3.8	God's Word	3.17	Examples:
3.9	God's Word		To control pollution, to develop natural resources
3.10	God		

SECTION ONE

1.1 Either order:
    a. takes up space
    b. has weight and mass

1.2 Examples:
    a. color, size, shape, taste,
       association, use, composition
    b. color, size, shape, taste,
       association

1.3 Examples:
    a. taste, temperature, smell, use
    b. taste, color, feel, carbonation
    c. taste, effect on boiling point
    d. color, texture, taste

1.4 Classmate check

1.5 mass per unit volume; amount of mass in a given amount of space

1.6 amount of space an object occupies or contains: l x w x h

1.7 Amount of matter in an object

1.8 centimeters (also milliliters)

1.9 grams

1.10 $1\frac{g}{cm^3}$

1.11 volume

1.12 If the mass of the amount of liquid displaced is equal to or greater than the mass that the object displaced, it will float.

1.13 The upward force of a liquid on an object.

1.14 displaces

1.15 Celsius

1.16 freezing

1.17 0°

1.18 32°

1.19 100

1.20 Hint:
Avoid substances which merely melt.

1.21 Both; chemical because of a change in the chemical composition, and physical because of change in the physical form

1.22 a. Chemical
    b. metal (substance)
    c. oxygen

1.23 water

1.24 ammonia, vinegar, lemon

1.25 yes, the water turned brown

1.26 brownish-red

1.27 Either order:
    a. definite shape
    b. definite volume

1.28 a. with a definite geometric shape (crystalline)
    b. Examples; snow, salt
    c. without definite geometric form (amorphous)
    d. Examples; glass, butter, wax

1.29 time varies

1.30 on the clip or the thread

1.31 cubic

1.32 a. mass
    b. space
    c. shape

1.33 farther

1.34 level

1.35 expands

1.36 Approximately 250 ml

1.37 The total volume will be less than 250 ml.

1.38 BB's fill the spaces between the marbles.

1.39 The air molecules inside the tire are heated and increased in temperature. An increase in temperature will result in an increase in pressure because the molecules move farther apart.

1.40 a. They are all matter: they have mass and take up space.
b. Each one differs in shape, state (solid, liquid, gas), density, color

1.41 matter

1.42 motion

1.43 pressure

1.44 plasma

1.45 stars

1.46 energy

SECTION TWO

2.1
Answers
may vary:

2.2 false

2.3 false

2.4 true

2.5 true

2.6 true

2.7 orbitals, rings, or shells

2.8 mass

2.9 nucleus

2.10 two

2.11 Either order:
a. protons
b. neutrons

2.12 b. valence energy level

2.13 c. 5, 6

2.14 c. valence

2.15 c. 1,800

2.16 b. special properties

2.17 b. quarks

2.18 sharing

2.19 electrons or protons

2.20 element

2.21 solid

2.22 Either order:
a. shape
b. volume

2.23 molecule

2.24 false

2.25 false

2.26 false

2.27 true

2.28 true

2.29 1

2.30 8

2.31 1

2.32 6

2.33 Inner (1)

2.34 H_2O

2.35 The 2 in the symbol H_2O means that each molecule of water contains two hydrogen atoms.

SECTION THREE

3.1 It became a liquid.

3.2 It went from white to clear.

3.3 yes – a black substance
3.4 carbon
3.5 hydrogen and oxygen
3.6 water
3.7 an element
3.8 atoms
3.9 three
3.10 one
3.11 100
3.12 true
3.13 true
3.14 true
3.15 true
3.16 false
3.17 atomic number
3.18 p + n

3.19 a nucleus

3.20 Ca
3.21 Carbon   Magnesium

3.22 atomic number = 79
     atomic mass = 197
     symbol = Au
     number of protons
        = 79
     electron distribution
        = 2, 8, 18, 32, 18, 1

3.23 atomic number = 26
     atomic mass = 56
     symbol = Fe
     number of protons
        = 26
     electron distribution
        = 2, 8, 14, 2

3.24 teacher check
3.25 With over a hundred different elements joined in different combinations, the number of compounds is very large. Also, varying the number of atoms adds to this list. Example: CO and CO_2 – one is carbon monoxide and the other is carbon dioxide.

3.26 a. Na (sodium)
     b. Cl (chlorine)
3.27 a. sodium - 1
     b. hydrogen - 1
     c. carbon - 1
     d. oxygen - 3
3.28 a. carbon
     b. hydrogen
     c. oxygen
     d. nitrogen
     e. sulfur
     f. iodine
     g. phosphorus
     h. calcium
     i. iron
3.29 You are cute.
3.30 ScIENCe IS GaROOVY
3.31 elements
3.32 iron
3.33 ammonia
3.34 Either order
     a. nitrogen
     b. hydrogen
3.35 oxygen
3.36 C

3.37   M
3.38   M
3.39   M

3.40   C

3.41   M

3.42   C
3.43   C

3.44   M
3.45   M

3.46   dissolved in the water

3.47   on the filter

3.48   by evaporating the water

SECTION ONE

1.1   A physical change occurred.

1.2   The appearance changed but not the composition.

1.3   Any four; any order:
   a. hardness
   b. density
   c. shape
   d. odor
      or
      taste, mass
      boiling point,
      freezing point, etc.

1.4   Any one; any order:
   freezing, melting, evaporation, dissolving, etc.

1.5   false

1.6   false

1.7   true

1.8   glass

1.9   oak wood

1.10   0.54 cm

1.11   0.08 cm

1.12   true

1.13   true

1.14   false

1.15   true

1.16   false

1.17   contract

1.18   mass

1.19   density or volume (size)

1.20   mass

1.21   density

1.22   a. It expanded due to heat.
       b. summer

1.23   It was left behind as a solid.

1.24   evaporation

1.25   Example:
       salt water

1.26   true

1.27   true

1.28   Example:
       -4° C

1.29   0° C

1.30   100° C

1.31   remained the same

1.32   It was either melting or vaporizing while absorbing the heat energy.

1.33   to produce a physical change

1.34   false

1.35   Any order:
   a. solid
   b. liquid
   c. gas

1.36   kinetic theory

1.37   condensation

1.38   evaporation

1.39   distillation

1.40   true

1.41   false

1.42   false

1.43   Heat was removed from the boiling water to melt the ice and to heat the melted water. The removed heat lowered the temperature of the water below the boiling point.

1.44   latent heat of vaporization

1.45   latent heat of fusion

1.46   specific heat

1.47   one with plaster of Paris

1.48   plaster of Paris

1.49   Heat is given off. Physical property changed.

1.50   A chemical change alters the substance to become a completely different and new substance. A physical change simply alters the original substance's physical state but leaves the substance chemically unchanged.

1.51   false

1.52   true

1.53   false

1.54   true

1.55   true

1.56   chemical property

1.57   chemical change

1.58   physical change

1.59   c. atoms

1.60   b. molecules

1.61   b. two

1.62   a. atoms

1.63   c. oxygen

1.64

1.65   a. Magnesium plus oxygen makes magnesium oxide.
       b. Iron plus oxygen makes iron oxide.

1.66   a. $2\,Mg + O_2 \rightarrow 2\,MgO$
       b. $4\,Fe + 3\,O_2 \rightarrow 2\,Fe_2O_3$

1.67   b

1.68   d

1.69   a

1.70   f

1.71   c

1.72   4

1.73   4

1.74   Conservation of Mass

1.75   2

1.76   2

1.77   10

1.78   8

1.79   5

1.80   7

1.81   Complete burning produces water vapor and carbon dioxide. Incomplete burning leaves some carbon soot unburned.

1.82   The candle stops burning when the oxygen is used up.

1.83   oxidation

1.84   carbon

1.85   condensation

1.86   chemical

1.87   heat

1.88   energy

1.89   93

1.90   radiation

1.91   radioactive

1.92   gamma

1.93 Either order:
      a. 2 protons
      b. 2 neutrons
1.94 a. an electron
1.95 b. physical
1.96 b. fission
1.97 c. chain reaction
1.98 b. energy
1.99 true
1.100 false

1.101 false
1.102 false
1.103 true
1.104 f
1.105 c
1.106 b
1.107 a
1.108 d

SECTION TWO

2.1   c
2.2   f
2.3   a
2.4   b
2.5   d
2.6   g
2.7   true
2.8   true
2.9   true
2.10 false
2.11 classmate check
2.12 A hydrogen ion is a hydrogen atom that has lost its electron; a proton.
2.13 An acid is defined by the production of H^+ ions, or H_3O^+ ions, in solution.
2.14 true
2.15 false
2.16 false
2.17 true
2.18 indicator

2.19 Either order:
      a. litmus paper
      b. phenolphthalein
2.20 a. blue
      b. red
2.21 colorless
2.22 hydrogen (H^+) or hydronium (H_3O^+)
2.23 lemon juice
2.24 milk, ammonia
2.25 yes
2.26 water
2.27 Any order:
      orange juice, vinegar, lemon juice, cider
2.28 because of the indicator turning from blue to red
2.29 Hydrochloric acid (HC1)
2.30 1.1
2.31 milk
2.32 7
2.33 b. H_3O^+
2.34 b. 0-14
2.35 a. 0-3

SECTION THREE

3.1   alkali or antacid

3.2   sodium hydroxide

3.3   Either order:

     a. fats

     b. oils

3.4   Any order:

     a. hydrogen

     b. oxygen

     c. metal

3.5   hydroxide

3.6   Either order:

     a. bitter

     b. slippery

3.7   e

3.8   d

3.9   b

3.10   c

3.11   a

3.12   hydroxide, or OH^-

3.13 OH^-

3.14   weak

3.15   Either order:

     a. hydrogen

     b. hydroxide

3.16   Either order:

     H^+, OH^-

3.17   two

3.18   hydroxide ion

3.19   hydrogen ion

3.20   pH

3.21   water

3.22 H_2O

3.23 H_3O^+

3.24   hydronium or acid

3.25   true

3.26   false

3.27   false

3.28   true

3.29   Any order

     milk of magnesia, soapy water, ammonia

3.30   turned pink

3.31   Any order:

     milk of magnesia, soapy water, ammonia

3.32   Any order:

     ammonia, salt water, milk of magnesia, baking soda

3.33   pink

3.34   teacher check

3.35   potassium hydroxide

3.36   distilled water

3.37   base

SECTION FOUR

4.1   when it became clear as water

4.2   pink

4.3   Because phenolphthalein is an indicator for bases and turns pink when a base is present.

4.4   Either order:

     a. acid

     b. base

4.5   salt

4.6   antacid

4.7   Either order; any two:

    a. leaves film

    b. requires more detergent

4.8   No. These substances are not NaCl and are chemically different and could be harmful.

4.9   false

4.10   true

4.11   false

4.12   true

4.13   teacher check

4.14   electrolytes

4.15   ions

4.16   positive

SECTION ONE

1.1   energy or quick energy

1.2   energy (stored), healthy skin,
       to carry vitamins A, D, E, and K

1.3   cell reproduction and repair,
       amino acids, form antibodies

1.4   cake, candy, cereal, potatoes,
       and other starchy, sweet foods

1.5   fish, poultry, eggs, milk, cheese,
       dried beans, peas, whole grains, etc.

1.6   e

1.7   k

1.8   b

1.9   g

1.10   d

1.11   f

1.12   c

1.13   j

1.14   h

1.15   a

1.16   i

1.17   Any order:
       carbohydrates, proteins, fats,
       minerals, vitamins, and water

1.18   a food containing carbon,
       hydrogen, and oxygen; includes the
       starches and sugars.

1.19   oily or greasy animal or
       vegetable substance.

1.20   a food containing carbon, hydrogen,
       oxygen and nitrogen in the
       form of complex amino acids.

1.21   any nonanimal, nonvegetable
       substance found in the earth,
       rocks, or water.

1.22   one of several complex organic
       substances necessary in small amounts
       in our diet

1.23   milk, cheese, dark green vegetables

1.24   milk, cheese, green vegetables,
       seafood, meats, eggs, cereals

1.25   dried beans, peas, whole grain
       cereals, liver, eggs, meat,
       green leafy vegetables, beets,
       raisins, apricots.

1.26   seafoods, iodized salt, some plants

1.27   milk and butter, vegetables,
       egg yolk, cheese, liver

1.28   fortified milk, butter, fish
       liver oil, sunshine

1.29   vegetable oils, wheat germ, liver,
       lettuce, whole grains

1.30   cheese, whole grains, green and
       leafy vegetables, peanuts, meat,
       fish, poultry, eggs, dried beans,
       milk

1.31   citrus fruits, most raw fruits
       and vegetables

Science 804 Answer Key

―――――――――――――

1.32   green vegetables

1.33   peas, whole grains, meat, fish poultry, eggs, milk, cheese, dried beans

1.34   bread, cake, candy, cereal, potatoes

1.35   peanuts, meat, butter, milk, cheese, salad dressings and milk

1.36   Sample Data:
   a. cracker; starch; tasted sweet; yes
   b. potato; starch; turned blue-black; yes
   c. cracker; starch; turned blue-black; yes
   d. bread; starch; turned blue-black; yes
   e. bread; sugar; no change in color; no
   f. bread; sugar; when saliva is added, a blue-black color; yes
   g. spinach; sugar; no color change; no
   h. spinach; mineral; whitish-gray ashes remain; yes
   i. bread; mineral; turns black, charcoal; no
   j. orange; mineral; burnt, charcoal color; no

   k. peanut butter; fat; toweling got greasy and translucent; yes
   l. butter; fat; toweling got greasy and translucent; yes
   m. egg yolk; fat; toweling got greasy and translucent; yes
   n. hair; protein; had a burned smell; yes
   o. meat, protein, had a burned smell; yes
   p. spinach; protein; turned color, little or no odor; no

1.37   a. breaking down
   b. changing (dissolving)
   c. bloodstream (plasma)

1.38   saliva

1.39   a. stomach
   b. enzymes (pepsin)
   c. acid (hydrochloric acid)

1.40   a. nutrients (particles)
   b. plasma (liquid)

1.41   mouth (saliva glands)

1.42   esophagus

1.43   liver

1.44   gall bladder

1.45   stomach

1.46   pancreas

1.47   large intestine

1.48   small intestine

SECTION TWO

2.1   no answer necessary

2.2   no answer necessary

2.3   no answer necessary

2.4   no answer necessary

2.5   carbohydrates

2.6   vitamins, minerals

2.7   protein, fats

2.8   sweets

2.9   Extras or Discretionary Calorie Allowance

2.10   teacher check for completeness and accuracy

2.11   teacher check for completeness, neatness, and thoroughness

2.12   teacher check for completeness, and accuracy of separation into the basic groups

2.13   teacher check for completeness, accuracy, and creativity of presentation.

SECTION THREE

3.1   answer depends upon survey results

3.2   teacher check for depth,-
       completeness, and accuracy

3.3   teacher check for reasonable
       group listing. Answer depends
       upon items (foods) listed in
       the previous survey.

3.4   Teacher check

3.5   Teacher check

3.6   Teacher check

3.7   d

3.8   g

3.9   a

3.10   e

3.11   c

3.12   b

3.13   f

3.14   teacher check for completeness
       and detail in the report.

3.15   teacher check for detail, specifics,
       and completeness of data.

SECTION FOUR

4.1   teacher check for detail, depth,
       and thoroughness.

4.2   antiseptic

4.3   improper sanitation

4.4.   a. epidermis
       b. dermis

4.5   acne

4.6   Answers will vary but should
       include frequent, thorough
       washing with soap and warm
       water, scrubbing, and rinsing
       with water.

4.7   Answers will vary but should in-
       clude brushing frequently, wash-
       ing, combining using mild sham-
       poo, and using a style and
       length that adds to your good
       grooming plan.

4.8   Any order:
       a. poor nutrition
       b. improper cleansing
       c. snacking without brushing

4.9   Answers will vary but should
       include brushing after every
       meal, rinsing, using dental
       floss, and periodic checkups.
       A good diet is also important.

4.10   Any order:
       a. proper diet– this provides
           the nutrients necessary to
           repair and maintain the body
           cells, bones, and teeth, and
           fight disease.
       b. cleanliness–prevents disease
           by reducing bacteria and germ
           contact with body cells; reduces
           growth places for microbes.
       c. regular checkups-prevents
           diseases, regular way to get
           help and advice on health
           problems.
       d. proper exercise–helps keep
           muscle tone, mental alertness,
           and remove body waste products.
       e. proper rest-necessary for cell
           repair and body rebuilding,
           especially in teenage years.
           (8-10 hours daily is advisable)

4.11   teacher check

SECTION ONE

1.1	push		1.19	no work
1.2	pull		1.20	work
1.3	push		1.21	b.
1.4	push		1.22	Answers will vary
1.5	pull		1.23	false
1.6	push		1.24	true
1.7	push		1.25	false
1.8	pull		1.26	kinetic
1.9	push		1.27	work
1.10	pull		1.28	potential
1.11	work		1.29	Either order: a. kinetic b. potential
1.12	work		1.30	kinetic
1.13	work		1.31	potential
1.14	work		1.32	kinetic
1.15	work		1.33	potential
1.16	no work			
1.17	work			
1.18	work			

SECTION TWO

2.1	heat		2.7	c. liquids and gases
2.2	friction		2.8	a. hot
2.3	true		2.9	true
2.4	false		2.10	false
2.5	false		2.11	false
2.6	a. convection current		2.12	true

2.13 true
2.14 b
2.15 d
2.16 e
2.17 c
2.18 A fluid is warmed. It weighs less, so it rises. Cool fluid (heavier) falls to take its place. This process sets up a circular convection current.
2.19 Air molecules are too far apart to cause their neighbors to vibrate.

2.20 Solar radiation is made up of waves similar to light waves. They do not need molecules to travel.
2.21 chemical reaction
2.22 oxidation reaction
2.23 Either order:
   a. combustion
   b. burning
2.24 fuel
2.25 Any order:
   a. coal
   b. oil
   c. wood
   d. natural gas
2.26 It was discovered that matter and energy could be changed into one another. The separate laws did not allow for this change, so a new law was stated.
2.27 Teacher check:
Look for depth, accuracy, and number of references used.
2.28 $E = mc^2$ Energy equals mass times the square of the speed of light.
2.29 Either order:
   a. nuclear reaction
   b. atomic reaction
2.30 true

2.31 true
2.32 true
2.33 false
2.34 true
2.35 false
2.36 false
2.37 false
2.38 false
2.39 anything which occupies space
2.40 the smallest part of an element that retains the characteristics of the element
2.41 neutron
2.42 proton
2.43 electron
2.44 nucleus
2.45 uranium - 235
2.46 proton (or neutron)
2.47 isotope
2.48 free neutron
2.49 unstable
2.50 chain reaction
2.51 energy
2.52 atom(ic) bomb
2.53 subdue
2.54 d
2.55 f
2.56 a
2.57 i
2.58 g

2.59   c

2.60   h

2.61   j

2.62   e

2.63   c. hydrogen, helium

2.64   c. thermonuclear

2.65   a. neutron

2.66   a. its high heat

SECTION THREE

3.1   God or Jesus Christ

3.2   mechanical

3.3   electrons

3.4   burning (or combustion or rapid oxidation)

3.5   radiant

3.6   water flowing over a dam

3.7   two magnets

3.8   false

3.9   false

3.10   true

3.11   true

3.12   false

3.13   a. mechanical
      b. mechanical
      c. mechanical
      d. mechanical
      e. mechanical
      f. chemical
      g. mechanical
      h. mechanical

3.14   a. mechanical
      b. mechanical
      c. mechanical
      d. mechanical

3.15   a. mechanical
      b. chemical
      c. heat
      d. mechanical

3.16   Either order:
      a. matter
      b. energy

3.17   organized

3.18   absorbed (lost)

3.19   entropy

3.20   entropy

3.21   Second Law of Thermodynamics

3.22   randomness (or disorder)

3.23   maximum entropy

3.24   God

3.25   a

3.26   e

3.27   b

3.28   f

3.29   c

SECTION ONE

1.1   approximately 950 miles
1.2   approximately 1,600 miles
1.3   south
1.4   from the bar magnet
1.5   c
1.6   d
1.7   false
1.8   false
1.9   false

1.10   true
1.11   Rome
1.12   William Gilbert

1.13   magnetite
1.14   decreases

1.15   declination
1.16   field
1.17   induced
1.18   Atoms are lined up in one
       direction in the metal.

1.19   Atoms in the bar are lined
       up (aligned) in one direction.
1.20   The atoms are disordered, which
       destroys the magnetism.

1.21   Either order:
       a. nickel
       b. cobalt
1.22   like poles repel
1.23   unlike poles attract
1.24   prevents them from clinging to
       the magnets

1.25   no
1.26   yes
1.27   poles
1.28   lines of force or field
1.29   domains
1.30   true
1.31   false
1.32   true
1.33   small region within a magnet– a
       concept useful in studying magnets
1.34   an alloy made of aluminum, nickel,
       and cobalt
1.35   the strength of a magnet decreases
       as the square of the distance
       from the source
1.36   one of two poles to which a com-
       pass will pass
1.37   a
1.38   d
1.39   The two N poles would move away
       from each other. If the magnets
       were on a frictionless surface,
       an N pole would link up with an
       S pole.
1.40   Oersted

1.41

compass

1.42   true
1.43   false

1.44   -273°

1.45   superconductor

1.46   electromagnet

1.47   increases

1.48   An iron bar is wrapped with wire. When electricity is run through the wire, the bar becomes a magnet.

1.49   Any order:
   a.   strength of the electric current
   b.   number of turns of wire
   c.   composition of the core

1.50   Supercold reduces electrical resistance, which increases the current which increases the magnetic strength.

SECTION TWO

2.1   They separated.

2.2   They collapsed.

2.3   They separated.

2.4   They touch.

2.5   no

2.6   The charge was shared when they touched.

2.7   Any order:
   a.   Objects with unlike charges attract each other
   b.   Objects with like charges repel each other
   c.   Charged objects attract neutral objects.

2.8   Greek

2.9   true

2.10 false

2.11 true

2.12 true

2.13 d

2.14 c

2.15 positively

2.16 ion

2.17 static

2.18 electroscope

2.19 charged

2.20 electrostatic (Van de Graff) generator

2.21 lightning rod

2.22 xerography

2.23 It has an equal number of protons and electrons.

2.24 Any order:
   a.   objects with unlike charges attract
   b.   objects with like charges repel
   c.   charged objects attract neutral objects

2.25 Examples:
   a.   lie flat
   b.   seek the lowest point available
   c.   avoid water
   d.   avoid explosive or flammable material

2.26 false

2.27 false

2.28 b

2.29 a

2.30 conductor

2.31 insulator

2.32 Any order:
   a.   silver
   b.   gold
   c.   copper
   d.   aluminum or iron

2.33 Any order:
a. type of material
b. thickness or length
c. temperature

2.34 Examples; Any order:
a. glass
b. porcelain
c. plastic
d. paper
e. rubber
f. air

2.35 teacher check
2.36 V = I x R = (0.3 amperes) (5 ohms)
V = 1.5 volts

2.37 dynamo
2.38 nichrome

2.39 branches
2.40 increase
2.41 circuit
2.42 current
2.43 d
2.44 series resistance: 12Ω + 10Ω
+8Ω = 30Ω

2.45 true

2.46 A fuse breaks (disconnects) a circuit in the event of a short circuit– an unexpected rise in current.

2.47 Either order:
a. series
b. parallel

2.48

2.49 a. volt
b. ampere
c. ohm

2.50 Either order:
a. metal plates
b. liquid (acid)

2.51 He caused an electrical spark to arc across a gap.

2.52 V = I x R = (3 amperes) (4 ohms)
V = 12 volts

2.53 V = I x R
24 volts = 1/2 ampere x R
48 ohms = R

2.54 Voltage in a circuit equals current times resistance or V = I x R.

2.55 proved that lightning is a form of static electricity

2.56 applied the inverse square law to electrical charges

2.57 made the first battery

2.58 predicted a relationship between electricity and magnetism

2.59 reported that two wires carrying a current exert forces on each other

2.60 found that the amount of current depended on the voltage

2.61 observed that a current in a wire can induce a current in another wire while the current is changing directions

2.62 announced the first working electric light bulb

2.63 built the first alternating current system in the United States

2.64 a
2.65 power
2.66 semiconductors
2.67 Rate of energy use =
power = volts x amperes
= 120v x 2 amperes
power = 240 watts

SECTION THREE

3.1   remote rural areas

3.2   22 percent

3.3   Either order:

    a. Rance River, France

    b. Bay of Fundy, Canada

3.4   Either order:

    a. tunnel

    b. strip

3.5   black lung or silicosis

3.6   Any order:

    a. impure air

    b. tunnel collapse

    c. black lung or silicosis

3.7   Any order; any four:

    a. magnetometers

    b. gravitational variation

    c. sound waves

    d. infrared film

    e. seismic waves

3.8   fuel for cars and trucks

3.9   teacher check

3.10   Examples; Any order:

    a. aspirin

    b. nylon

    c. dyes

    d. plastics

3.11   a. light

    b. electricity

3.12   satellite

3.13   hot springs

3.14   dissolved minerals

3.15   Any order:

    a. uranium

    b. plutonium

    c. thorium

3.16   Any order:

    a. no pollution

    b. short transmission lines

    c. can serve many people

    d. excellent safety record

SECTION ONE

1.1   a

1.2   c

1.3   b

1.4   a

1.5   a

1.6   a

1.7   b

1.8   yes

1.9   any one of them, by definition of cubit

1.10   Difference could range between a fraction of a shortest cubit and two or three shortest cubits.

1.11   String stretches and is therefore unreliable.

1.12   Use an average cubit.

1.13   Carry a strip of paper that is the length of your classroom cubit

1.14   count

1.15   do not count

1.16   do not count

1.17   count

1.18   do not count

1.19   do not count

1.20   do not count

1.21   count

1.22   by pairing them off and seeing which were left over

1.23   by comparing the amount of milk in two identical pails

1.24   by showing him your right shoelace, or by showing him something that is the same length as your right shoelace

1.25   B is longer

1.26   by matching each to another object – a pencil or piece of paper

1.27   teacher check

1.28   teacher check

1.29   teacher check

1.30   Hint:
difficult to measure; maybe $1/10$ pac

1.31   The pac is inconvenient for long distance or lengths shorter than one pac.

1.32   Fold the cover in half, then fold the half in half and so on to produce several equal minipacs.

1.33   The inch is conventionally divided into $1/2$'s, $1/4$'s, $1/8$'s. $1/16$'s and so on.

1.34   14.5 or 14 $1/2$

1.35   84

1.36   9.67 or 9 $2/3$ or $9.6\overline{6}$

1.37   348

1.38   15,840

1.39   1.89

1.40   3,333.3 or 3,333 $1/3$

1.41   120,000

1.42   237

1.43   300

1.44 Hint:
     compare with answer 1.57
1.45 Example:
     1.41 and 1.43
1.46 Example:
     They required multiplying by a
     power of 10; i.e., by 10,000 and
     by 100.

1.47 1.74
1.48 700

1.49 290

1.50 0.29
1.51 3,000
1.52 10

1.53 1,000,000
1.54 10,000,000
1.55 7.9
1.56 1

1.57 Example:
     time will be shorter than 1.44
1.58 The metric system requires multiplying
     or dividing by powers of 10.
1.59 teacher check
1.60 a. 23
     b. 1.6

    c. 0.5
    d. $3/4$
    e. 1 $3/5$
    f. 9
1.61 a. 4
    b. 10
    c. 21
    d. .9
    e. 1.5
    f. .04
1.62 a. 10
    b. 25
    c. 50
    d. 22.5
    e. 4
    f. 0.5
1.63 90°
1.64 by folding the corner of the page
     in half
1.65 20 m
1.66 by folding the top edge of a sheet
     of paper against the side edge at
     the corner, that is, by halving a
     90° angle
1.67 by halving a 45° angle; that is, by
     folding the paper at its corner a
     second time
1.68 approximately 600 m
1.69 approximately 24 m

SECTION TWO

2.1 force
2.2 vector quantities
2.3 gravity
2.4 newton

2.5 church
2.6 false

2.7 false

2.8 true

2.9 false
2.10 teacher check
2.11– 2.13 Examples:
2.11 An object needs to be pushed or
     pulled in order to change its
     motion or its position.
2.12 A big force causes a big change in
     motion; a small force causes a
     small change. As the mass increases,
     the force needed to accelerate
     must increase also.

2.13 Each time a force acts in one direction, an equal force acts in the opposite direction.

2.14 teacher check

2.15 will depend on surface

2.16 will depend on surface

2.17 The starting force is greater.

2.18 It is greater.

2.19 true

2.20 true

2.21 true

2.22 false

2.23 false

2.24

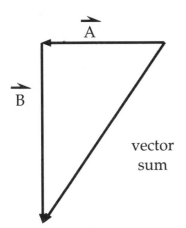

vector sum

2.25

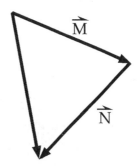

2.26

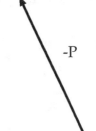

-P

SECTION THREE

3.1 false

3.2 true

3.3 true

3.4 true

3.5 true

3.6 false

3.7 work = force x distance
= 5 newtons x 1 meter
work = 5 joules

3.8 work = force x distance
work = 10 N x 3 m = 30 J

3.9 work = force x distance
work = 50 N x 2 m = 100 J

3.10 No work is done, because the force (the weight of the book) does not have the same direction as the distance, which is horizontal.

3.11 a. work = force x distance
= 500 lbs. x 12 feet
work = 6,000 ft. lbs.
b. increase in PE = work done on the pile driver = 6,000 ft. lbs.
c. The instant before the pile driver hits, its kinetic energy equals its maximum potential energy KE (bottom) = PE (top) = 6,000 ft. lbs.

3.12 a. work = force x distance
work = 4 N x 3 M
work = 12 J
b. PE equals lifting work
PE = work = 12 J

3.13 a. work = force x distance
work = 2 N x 3 m
work = 6 J
b. No change in height produces no change in PE.
c. If the trunk moved at constant speed, the force needed to move it is the force needed to overcome the frictional force. Frictional force, therefore, is 2 N.

3.14 power = $\frac{\text{force x distance}}{\text{time}}$

= $\frac{2,000 \text{ lbs x } 100 \text{ ft.}}{20 \text{ sec.}}$

power = 10,000 $\frac{\text{ft. lbs}}{\text{sec.}}$

or $\frac{10,000}{550}$ = 18.2 HP

NOTE: at 10 ft. per story, the distance between the first and eleventh stories is 100 ft.

3.15 Example:
a. Sam
b. 100
c. 10
d. 4
e. 250
f. 0.45

3.16 Hint:
The lighter one may be able to produce more power, although the opposite may also be true.

3.17 Hint:
The better the condition, the greater the power output.

3.18 probably not for very long

SECTION ONE

1.1	a		1.18	sliding
1.2	c		1.19	rolling
1.3	friction		1.20	fluid
1.4	parallel		1.21	teacher check
1.5	D		1.22	after the soap was rubbed on the block
1.6	U		1.23	lubrication
1.7	D		1.24	shape
1.8	U		1.25	friction
1.9	varies with the kind of coin used		1.26	Both preparations lessen friction. The cap lessens fluid friction with a smooth surface. The grease lessens fluid friction by lubrication.
1.10	varies with the kind of coin used			
1.11	should be less			
1.12	The rollers decrease the friction.			
1.13	true			
1.14	false		1.27	$\mu = \dfrac{F_f}{F_n} = \dfrac{9N}{18N} = 0.5$
1.15	true			
1.16	true			
1.17	starting			

SECTION TWO

2.1 closer

2.2 a. 4 kilograms

b. $4 \text{ kg} \times 9.8 \dfrac{m}{\text{sec.}} = 39.2 \text{ N}$

2.3 a. $\dfrac{6 \text{ meters}}{3 \text{ meters}} = 2$

b. $\dfrac{6,000 \text{ N}}{4,000 \text{ N}} = 1.5$

c. $\dfrac{1.5}{2.0} \times 100\% = 75\%$

2.4 false

2.5 false

2.6 true

2.7 true

2.8 false

2.9 $(?) (0.6 \text{ m}) = (3N) (0.4 \text{ m})$

$= \dfrac{(3N) (0.4 \text{ m})}{0.6 \text{ m}}$

$= 2 \text{ N}$

2.10 true

2.11 false

2.12 true

2.13 first

2.14 second

2.15 third

2.16 first

2.17 first

SECTION THREE

3.1	effort		3.8	b
3.2	true		3.9	d
3.3	true		3.10	a
3.4	false		3.11	4
3.5	5		3.12	friction
3.6	a. wheel		3.13	20
	b. decreasing		3.14	The blades went around more times than the handle.
3.7	direction			

3.15 3

3.16 a. Mark a point on the tire with chalk, thread, etc.

b. Rotate the pedal wheel through one rotation while watching the mark on the tire.

3.17 motion (direction)

3.18 Either order:
a. speed
b. direction

3.19 counterclockwise

3.20 clockwise

SECTION FOUR

4.1 books without the ruler

4.2 effort (force)

4.3 effort (force)

4.4 a. small
b. long

4.5 a. large
b. short

4.6 length

4.7 slope

4.8 a.

$$AMA = \frac{F_r}{F_e} = \frac{800\ N}{400\ N} = 2$$

b.

$$IMA = \frac{d_e}{d_r} = \frac{5}{2} = 2\frac{1}{2}$$

c. $efficiency = \dfrac{AMA}{IMA} \times 100\%$

$$= \frac{2}{2\,1/2} = 80\%$$

4.9 thinner

4.10 friction

4.11 false

4.12 true

4.13 Examples:
shelves
refrigerator feet
washer feet

furniture

turnbuckle

car jack

corkscrew

bolts (machinery)
spiral staircase
clamps

SECTION ONE

1.1   teacher check

1.2   teacher check

1.3   reduce water loss

1.4   chloroplasts

1.5   Any order:
   a. green
   b. yellow
   c. orange
   d. red

1.6   Moisture collects on the plastic bag.

1.7   transpiration

1.8   through the stomata

1.9   little or no moisture

1.10   They must survive in the desert where there is little water.

1.11   c

1.12   f

1.13   e

1.14   d

1.15   g
1.16   b
1.17   a
1.18   mostly on underside of leaves
1.19   guard cells
1.20   They may die.
1.21   They allow CO_2 to enter and water and oxygen to escape.
1.22   They have no leaves to lose water.

1.23   Any order:
   a. nitrogen: green foliage and rapid growth
   b. phosphorus: strong roots and stems
   c. potassium: disease resistance

1.24   transpiration

1.25   trace elements

1.26   chlorophyll
1.27   large surface area

1.28   stomata
1.29   a. root
   b. leaf
   c. flower
   d. seed
   e. stem
   f. root
   g. flower
   h. seed
   i. fruit
   j. seed
   k. stem
   l. leaf
   m. tuber
   n. root
   o. stem
   p. flower
   q. seed
   r. tuber
1.30   teacher check
1.31   d
1.32   h
1.33   g
1.34   e
1.35   l
1.36   a

1.37 j

1.38 c

1.39 b

1.40 k

1.41 i

1.42 $6CO_2$, $12H_2O$, light, chlorophyll, $6O_2$, $6H_2O$

1.43 carbon dioxide

1.44 light

1.45 They provide carbon dioxide.

1.46 animal respiration

1.47 They would eventually die because they need the CO_2 from animal respiration.

1.48 Either order:
   a. They would starve because they need plants as food.
   b. They would die from lack of oxygen

1.49 to 1.55 Examples:

1.49 a. wide variety of food, lots of meat, fruits, vegetables, nutrition very important
   b. some days no food at all, dried berries, roots, maybe raw meat if lucky

1.50 a. garments for every temperature and activity, special shoes for hiking, gym, dress up
   b. little clothing, animal skins

1.51 a. almost no limit if a person wishes to learn
   b. none, learn food gathering from watching elders

1.52 a. radio, games, skating, hiking, camping, clubs
   b. no time for play, most of the day spent looking for food

1.53 a. limitless, bikes, clothing, games, books, radios, clocks
   b. few, if any, perhaps fish hook or basket

1.54 a. family car, bicycle, train, airplane, bus, boat
   b. walked everywhere

1.55 a. excellent medical care: medicine, shots, dental work if needed, hospitals
   b. no medical or dental care; probably malnourished

1.56 Examples:
   a. Shasta daisy
   b. Burbank potato
   c. Santa Rosa plum
   d. thornless prickly pear
   e. elephant garlic

1.57 teacher check

1.58 Any order:
   a. greater production
   b. better quality
   c. special characteristics

1.59 Examples; either order:
   a. disease resistance
   b. earlier ripening

1.60 Austria

1.61 1860's

1.62 laws of inheritance

1.63 it had no pit

1.64 a cross between two known but unlike strains of plant or animal

1.65 The hybrid will either cross with itself or with an unknown plant.

1.66 Wheat had to be harvested quickly.

1.67 Machines allowed harvesting to be faster. Hybrid wheat did not fall out so easily.

1.68 increasing

1.69 staying about the same

1.70 2

1.71 fertilize

1.72 Either order:
   a. wells
   b. irrigation

1.73   It is a source of protein.
1.74   Any order:
    a. bloated belly
    b. thin arms and legs
    c. discolored hair
    d. mental retardation
1.75   one-third
1.76   Example:
With population increasing and food supplies staying level less food is available per person each year. Since distribution is not equal, some people will have enough but others will starve.

1.77   teacher check
1.78   a. 8
    b. 20
    c. 35
1.79   Either order:
    a. starch
    b. sugar
1.80   Any order:
    a. calories
    b. vitamins
    c. minerals
1.81   protein

SECTION TWO

2.1   It is not a form the plant can absorb.

2.2   They provide nitrogen compounds.
2.3   They provide food.
2.4   nitrogen in the air between the soil particles

2.5   They are excreted into the surrounding soil.

2.6   two-thirds
2.7   converts them into proteins particular to itself

2.8   a. Jefferson
    b. It increases soil fertility.
2.9   plowed-under legume crops
2.10   Examples; any order:
    a. beans
    b. peas
    c. clover
    d. peanuts
2.11   teacher check
2.12   Either order:
    a. bacteria
    b. fungi

2.13   Their decomposition puts valuable elements back into the environment so they can be used again.
2.14   Life would cease; we would run out of elements.
2.15   They have no chlorophyll.
2.16   Any order:
    a. drying, or salting
    b. cooking
    c. canning
    d. freezing
    e. refrigerating
2.17   Any order:
    a. cheese
    b. penicillin
    c. bread yeast
    d. edible mushrooms
2.18   Either order:
    a. elements are released for reuse
    b. the amount of organic matter is reduced
2.19   simple plants that lack chlorophyll and derive their food from other organisms
2.20   teacher check
2.21   helper check

2.22   snow, rain, hail

2.23   temperature, humidity, wind,
        surface area

2.24   respiration, excretion, perspiration

2.25   teacher check

2.26   true

2.27   false

2.28   false

2.29   false

2.30   false

2.31   false

2.32   true

2.33   false

2.34   false

2.35   true

2.36   false

2.37   Photosynthesis
        a. CO_2
        b. O_2
        c. sun
        d. food manufactured; glucose

        Respiration
        a. O_2
        b. CO_2
        c. food
        d. food used; energy used

SECTION THREE

3.1   a plant (a food-making organism)

3.2   any animal (food-consuming organism)

3.3   an animal that eats only plants

3.4   an animal that eats only other animals

3.5   an animal that feeds on dead material

3.6   an animal that eats both plants
        and animals

3.7   conditions that regulate the
        number of any species

3.8   an organism that breaks down
        organic matter

3.9   an animal that is not eaten by
        any other animal

3.10   a specific place where an organism
        naturally occurs

3.11   a group of animals and plants
        living together

3.12   the study of organisms in
        relation to their environment

3.13   The owl population would increase.

3.14   Their numbers would drop.

3.15   The plants would die so both mouse
        and owl populations would drop.

3.16   Energy and biomass are lost at
        each step.

3.17   a. scavenger
        b. producer
        c. herbivore
        d. decomposer
        e. herbivore
        f. decomposer
        g. carnivore
        h. producer
        i. scavenger
        j. decomposer
        k. carnivore
        l. scavenger (carnivore)
        m. herbivore
        n. producer
        o. carnivore

3.18 teacher check
3.19 Either order:
a. disease
b. famine
3.20 Any order:
a. Industrial Revolution
b. medicine
c. agriculture
3.21 Mating and care of the young are disrupted and the population drops.
3.22 It does not easily decompose.
3.23 an organism that is not killed by a poison designed to kill it

3.24 Any order
a. dysentery and eye disease
b. malaria and yellow fever
c. the plague
3.25 It disrupts the reproductive cycle so no eggs are laid or so the ones that are laid have thin shells.
3.26 teacher check
3.27 teacher check
3.28 teacher check
3.29 a. ivory
b. hide for purses and shoes
c. skin for fur coats
d. sold as potted plants
e. trophy animals
f. feathers and pets
3.30 Japanese beetles were a natural control for corn borers.
3.31 Examples: Any three, any order:
a better clothing
b. fast transportation
c. comfortable homes
d. nutritious food
e. food production
f. medicine
3.32 Examples; Any order:
a. people poisoned or made mentally retarded
b. some chemicals are too poisonous to keep around
c. kills natural controls

3.33 Examples; Any order:
a. ignorance-killing natural controls
b. greed-killing of rare animals for fur
c. population-trash
d. technology-chemicals that pollute; automobiles
3.34 teacher check
3.35 Either order:
a. productive topsoil is being eroded
b. soil bacteria are being killed by chemicals
3.36 The land is bare from fall harvest until spring planting.
3.37 Examples: Any five, any order:
a. nature study
b. camping
c. hiking
d. lumber
e. paper
f. food (hunting)
g. habitat for animals
3.38 Any order:
a. aluminum
b. iron
c. copper
3.39 Any order:
a. coal
b. oil
c. gas
3.40 remains of plants and animals
3.41 Either order:
a. plastic
b. energy
3.42 Examples; Any order:
a. sun– expense, northern days are shorter in winter
b. wind– not enough wind, wind mills use metal
c. nuclear– danger of accidents, cooling problems, nuclear waste
3.43 Either order:
a. It is becoming polluted.
b. It is being used up 140 times faster than it is returning.

3.44   60 percent

3.45   Either order:

a. They are the food producers for the entire ocean.

b. They provide 50% of the oxygen given off by photosynthesis.

3.46   Either order:

a. uses oxygen

b. gives off pollutants

3.47   they bring enjoyment

3.48   teacher check

3.49   teacher check

3.50   Examples; Any order:

a. soil– proper plowing

b. forests– recycle paper

c. minerals– recycle minerals

d. fossil fuel– walk more

e. air– use less fuel

f. water– shorter showers

g. wildlife– provide forests for them

h. wilderness– purchase more areas

SECTION ONE

1.1   a. choose a problem

       b. state what you think is the probable solution (hypothesis)

       c. research what other scientists have done to solve your problem

       d. experiment to prove or disprove your hypothesis

       e. state the hypothesis again as a theory

       f. if wrong, state a new hypothesis

       g. write a paper on what you did to prove your hypothesis

       h. change your hypothesis if it is proved wrong

       i. state the theory as a law

1.2   c

1.3   a

1.4   a

1.5   b

1.6   c

1.7   b

1.8   c

1.9   a. Investigation #1

       b. Problem:
Has science advanced since Aristotle? How?

       c. Materials:
Science LIFEPAC 801, an encyclopedia, or online resources

       d. Method:
Read *A BRIEF HISTORY OF SCIENCE* from Science LIFEPAC 801, or an article about the history of science in an encyclopedia or online resource.

       e. Result:
Many people contributed to the advancement of science.

       f. Answers may vary:
Aristotle studied nature and wrote his ideas in an orderly manner. He did not experiment, therefore, his conclusions were often wrong. Democritus said the smallest piece of matter was an atom. Atomic science is based on his idea. In the Middle Ages alchemists made discoveries as they tried and failed to turn metals into gold. Men began sorting out facts and writing them down. Men such as Galileo published

several scientific papers, and Sir Isaac Newton added his mathematical findings to the process. The biological sciences advanced with La Marck. In modern science Curie and Einstein made important discoveries.

g. Conclusion; Example: Science has advanced since Aristotle's day because people observed, investigated, experimented, and kept records.

1.10 kilo, kg

1.11 hecto, hg

1.12 deca

1.13 none, g

1.14 deci, dg

1.15 centi, cg

1.16 milli, mg

1.17 kilometer = 1,000 meters, km

1.18 hectometer = 100 meters, hm

1.19 dekameter = 10 meters, dkm

1.20 decimeter = $\frac{1}{10\text{th}}$ meter, dm

1.21 centimeter = $\frac{1}{100\text{th}}$ meter, cm

1.22 millimeter = $\frac{1}{1000\text{th}}$ meter, mm

1.23 Any order:
   a. Everyone will have to learn the metric system.
   b. Schools will have to have courses for teaching metrics.
   c. Metric tools for measuring will have to be bought.
   d. Tools used to build things (autos) will have to be changed to metrics.
   e. Everything will have to be relabeled or keyboards

may have to be changed; conversion will cost money; prices will go up.

1.24 a. 88 km/h
     b. 55 mph

1.25 density

1.26 buoyancy

1.27 Celsius

1.28 bottom

1.29 odor, taste, color, density, brittleness, hardness, luster, form, buoyancy

1.30 yes

1.31 Times will vary depending on room condition.

1.32 time will vary

1.33 e

1.34 f

1.35 b

1.36 h

1.37 g

1.38 c

1.39 a

1.40 H–O–H

1.41 O–Si–O

1.42 Na—Cl

1.43 a. Niels Bohr
     b. orbits

1.44 atoms

1.45 symbols

1.46 nuclear energy

1.47 a. 3
      b. 1
      c. 1
1.48 any number from 1 to 106
1.49 mixture
1.50 separated by student with toothpicks
1.51 Any order:
      a. iron filings
      b. sand
      c. salt
      d. sawdust
1.52 Examples:
      a magnet to remove iron filings, add water so sawdust would float to top to be skimmed off, saltwater poured off to leave sand, saltwater poured off to evaporate water leaving salt
1.53 to record the pressure as it increases and to prevent damage if too much pressure builds up
1.54 to keep steam from escaping
1.55 The water changes phases from a liquid to a gaseous state. The food cooks faster with the increased temperature.

1.56 At high altitudes the air pressure is reduced. Regular directions are for sea level.
1.57 $4\ Fe + 3\ O_2 \longrightarrow 2Fe_2O_3$
1.58 a. hydrogen
      b.– c. Either order:
      b. metal
      c. hydroxide
1.59 salt
1.60 b
1.61 a
1.62 b
1.63 a
1.64 c
1.65 c
1.66 a
1.67 b

SECTION TWO

2.1 a. kinetic
     b. potential
     c. kinetic
     d. kinetic
     e. kinetic
     f. potential
2.2 conduction
2.3 convection
2.4 a. radiation
     b. light
2.5 a. expand
     b. contract

2.6 water cooled to ice
2.7 But even the very hairs of your head are all numbered. Fear not therefore: ye are of more value than many sparrows.
2.8 a. 1
     b. 2
     c. 1
2.9 a. 2   c. 1
     b. 4   d. 2

2.10   a. 1
      b. 2
      c. 2

2.11   all fall over in a chain reaction

2.12   When the first atom splits because the nucleus has been bombarded, the parts hit the next or close atoms and they split, and on and on.

2.13   teacher check

2.14   energy

2.15   radioactive

2.16   entropy

2.17   God

2.18   Any order:
      a. Magnetism can be turned on or off.
      b. The strength of the magnet can be changed.
      c. The poles can be reversed.

2.19   Either order:
      a. to separate iron from other materials
      b. to hold items on bulletin boards, refrigerators, oven doors etc.

2.20   Any order:
      a. for lifting scrap iron, iron metal, and so forth
      b. telephone
      c. telegraph
      or electronic motors; generators

2.21   a. cloud (negative charge)
      b. leader
      c. lightning (return stroke)
      d. earth (positive charge)

2.22   Either order:
      a. like charges repel
      b. unlike charges attract

2.23   It moves across space to the opposite charge because opposites attract.

2.24   Any order:
      a. do not stand under a tree
      b. do not stand on metal or hold metal
      c. keep out of the open – lie down in a low spot

2.25   a. yes
      b. Tires act as insulators.

2.26   Examples:
      a. City transportation stopped; people could not get home.
      b. Refrigerators and freezers stopped; food spoiled.
      c. Hospitals operated at reduced power; patients were inconvenienced.
      d. Computers stopped; reservation systems were jammed.
      e. Water pumps stopped working; no water was available.
      f. Elevators stopped; people were trapped in them for hours.

2.27   Examples:
      a. Electric milking machines stopped.
      b. Equipment used to cool produce - - eggs, vegetables, milk - - stopped.
      c. Water pumps that supply water to livestock stopped.

2.28   Hint:
      Paragraph should indicate that family members discussed this topic with student. Plan made should indicate adult input and a practical solution. This assignment provides an opportunity for family participation.

2.29   built new type high-voltage generator used to smash atoms

2.30   found magnet could be turned on and off with switch-on current

2.31   discovered magnet moved near a coil causes electric current to flow in wire

2.32   made first electromagnet

2.33 found that electric current through a wire makes wire act like a magnet

2.34 chemical action could produce continuous flow of electricity

2.35 identified circumstances leading to production of current electricity

2.36 teacher check

2.37 Example:

$^1/_2$ for two-foot board

2.38 the amount of the advantage in using the machine

2.39 3 for 3' board

2 for 2' board

2.40 it took $^1/_2$ ($^1/_3$) the effort to raise it one foot

2.41 oil

2.42 vaseline or soap

2.43 soap or wax

2.44 grease

2.45 oil

2.46 Hint:

This story may be highly imaginative but should show that the pupil understands the part played by friction in everyday happenings.

2.47 true

2.48 false

2.49 false

2.50 true

2.51 true

2.52 a.

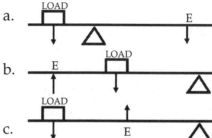

b.

c.

2.53

2.54 a.

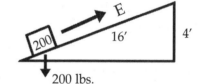

b. Effort x 16 = 200 x 4

Effort = 50

16 ÷ 4 = M.A. = 4

Effort = 200 ÷ M.A.

= 200 ÷ 4

= 50

SECTION THREE

3.1 two-thirds

3.2 chemically

3.3 enzymes

3.4 peristalsis

3.5 autonomic nervous

113

3.6 Examples:
roast beef, cereal, potato,
lettuce, milk, banana, butter,
peanuts

3.7 Examples:
potato chips, candy bar,
soda pop

3.8 Examples:
a. no
b. I had too much junk
food to eat.

3.9 Do not eat them. Try to learn
what foods to avoid.

3.10 a. Nutrition can be ignored.
b. Example:
Original ideas should be
accepted but an additive
should meet all the requirements
rather than any one.

3.11 Know what foods are in each
nutrition group and should eat a
well-balanced diet.

3.12 g
3.13 f
3.14 e
3.15 i
3.16 a
3.17 b
3.18 c
3.19 d

3.20 six molecules of carbon dioxide
plus 12 molecules of water produces
one molecule of sugar plus 6
molecules of water and 6 molecules
of oxygen gas

3.21 Legume Plant, Food with N,
N, Animal Waste, Decay, Earth

The Nitrogen Cycle

3.22 Evaporation, Storm Clouds,
Precipitation (rain, snow...)
Earth

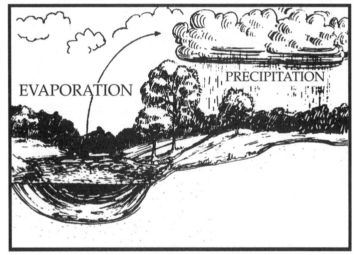

The Water Cycle

3.23 Animals, CO_2, O_2, Plants

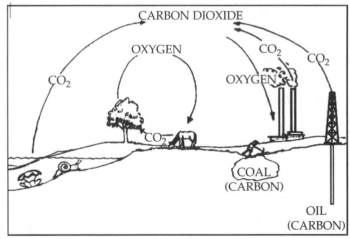

The Carbon-Oxygen Cycle

3.24 Plants Use Elements, Food,
Animals, Decay and Waste,
Elements in Soil

The Decay Cycle

3.25 created by God (natural)
3.26 balance
3.27 humans
3.28 Either order:
a. conservation
b. passing laws
3.29 teacher check

SECTION FOUR

4.1 teacher check
4.2 technology (science)
4.3 Either order:
a. idle
b. lazy
4.4 rewards or gifts
4.5 Examples:
auto mechanic, computer
programmer, dental hygienist,
geologist, practical nurse,
pharmacist, physician, psychologist,
scientific clerk
4.6 And also that every man should
eat and drink, and enjoy the
good of all his labor, it is the
gift of God.

4.7 keeping one seated much of the
time
4.8 good character
4.9 personality tests
4.10 tests that measure general
reasoning, form perception,
vocabulary, and so on
4.11 A job is a good experience to
teach values and responsibility.
4.12 teacher check

Notes

SELF TEST 1

1.01   knowledge

1.02   experimentation

1.03   Aristotle

1.04   gold

1.05   hypothesis or theory

1.06   Renaissance

1.07   earth

1.08   Sir Isaac Newton

1.09   Charles Darwin

1.010 microorganisms or organisms

1.011 b

1.012 f

1.013 d

1.014 a

1.015 c

1.016 c

1.017 b

1.018 a

1.019 a

1.020 c

1.021 milliliters

1.022 4.142×10^3

1.023 5,200

1.024 29.6

1.025 4

SELF TEST 2

2.01   significant figures

2.02   technology

2.03   tensile strength

2.04   wheel

2.05   crossbow

2.06   Democritus

2.07   metric

2.08   dynamo (generator)

2.09   Leeuwenhoek

2.010 communication

2.011 true

2.012 true

2.013 false

2.014 false

2.015 false

2.016 true

2.017 false

2.018 true

2.019 true

2.020 false

2.021 c. "animalcules"

2.022 a. irrigation

2.023 a. 5

2.024 c. radiation

2.025 c. gunpowder

SELF TEST 3

3.01   cancer

3.02   pollution-control

3.03   imperfect

3.04   fuels

3.05   cloning

3.06   space

3.07   shaduf

3.08 $8.24 \cdot 10^2$

3.09   Greek

3.010 $E = mc^2$

3.011   God's Word

3.012   Choose a problem

3.013   Make a hypothesis.

3.014   Research what others have done.

3.015   Perform experiments.

3.016   If true, restate the hypothesis as a theory.

3.017   If false, make a new hypothesis and begin again.

3.018   Write and publish a paper.

3.019   Change the theory if it should be proved wrong.

3.020   Restate the theory as a law.

3.021   e

3.022   f

3.023   b

3.024   d

3.025   c

SELF TEST 1

1.01–1.010
    x rice; x sugar; x wood; x milk;
    x air

1.011 j
1.012 f
1.013 d

1.014 a
1.015 b, c, e, or h
1.016 g

1.017 e

1.018 l
1.019 n, e, or c
1.020 i
1.021 o

Either order: 1.022–1.023
1.022 takes up space
1.023 has mass

Examples; Any order:
1.024 shape
1.025 color
1.026 odor
1.027 density
1.028 taste, buoyancy

1.029 a, c
1.030 b, c

1.031 b, d
1.032 2 cm x 4 cm x 3 cm = 24 cm^3

1.033 density $= \dfrac{mass}{volume} = \dfrac{112 \text{ g}}{24 \text{ cm}^3} = 4.7\ \dfrac{\text{g}}{\text{cm}^3}$

SELF TEST 2

2.01   h
2.02   m
2.03   k
2.04   j
2.05   g
2.06   f
2.07   e
2.08   b
2.09   c
2.010 d
2.011 true
2.012 false
2.013 false

2.014 true
2.015 true
2.016 false
2.017 true
2.018 true
2.019 false
2.020 false
2.021 2
2.022 2
2.023 8
2.024 6

2.025 8
2.026 b. proton
2.027 c. solid
2.028 c. nucleus
2.029 a. three atoms
2.030 b. seven
2.031

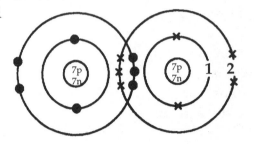

2.032 chemically combined (joined)
2.033 neutron
2.034 buoyancy
2.035 nucleus
2.036 atomic theory

SELF TEST 3

3.01   true
3.02   false
3.03   false
3.04   true
3.05   true
3.06   false
3.07   true
3.08   true
3.09   false
3.010 true
3.011 valence
3.012 electrons

3.013 1
3.014 volume
3.015 symbol

3.016 negative

3.017 two
3.018 elements or atoms
3.019 protons
3.020 nucleus
3.021 a. atomic number
3.022 b. symbol
3.023 c. name of element
3.024 d. atomic mass
3.025 e. electrons distribution
3.026

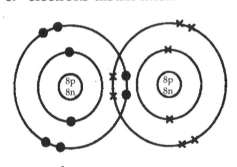

● = electrons
x = electrons

3.027 a. has mass and takes up space; fixed shape and volume; may be crystalline or amorphous

b. has mass and takes up space; fixed volume; no fixed shape

c. has mass and takes up space; neither fixed volume nor fixed shape.

3.028 A chemically combined substance made up of two or more different elements.

3.029 Substances that have only one kind of atom.

3.030 A reversible combination of two or more substances not in fixed proportions.

3.031 Either order:
a. has mass
b. occupies space

3.032 Any order; any three: density, buoyancy, freezing point/boiling point, color, taste, mass, size, shape

3.033 g

3.034 b

3.035 i

3.036 c

3.037 d

SELF TEST 1

1.01 h

1.02 k

1.03 c

1.04 m

1.05 e

1.06 g

1.07 b

1.08 a

1.09 j

1.010 d

1.011 i

1.012 l

1.013 C

1.014 P

1.015 P

1.016 P

1.017 P

1.018 C

1.019 P

1.020 A chemical change involves a change in composition of the matter. New substances are formed.

1.021 A change in appearance or physical property. No change in composition or molecular structure.

1.022 the way a substance reacts with another substance

1.023 Any order:
a. hardness
b. shape
c. color
d. taste
e. density
or size (volume), mass, melting and freezing points, etc.

1.024 Any order:

a. alpha particles
b. beta particles
c. gamma rays

1.025 Amount of heat needed to cause a solid to become a liquid after reaching the melting point.

1.026 Amount of heat needed to cause a liquid to become a gas after it has reached its boiling point.

1.027 Particles of a substance are constantly in motion.

1.028 a. Fe
b. O_2

1.029 10

1.030 10

1.031 Law of Conservation of Mass

SELF TEST 2

2.01 hydrogen

2.02 Either order:
a. proton
b. electron

2.03 fission

2.04 evaporation

2.05 ion

2.06   Any order:
     a. hardness
     b. shape
     c. taste
     d. color
     e. density
       or size (volume), mass,
       melting and freezing
       points, etc.

2.07   c
2.08   k
2.09   i
2.010 j
2.011 g
2.012 a
2.013 b
2.014 d
2.015 e
2.016 m

2.017 l
2.018 h
2.019 A
2.020 A
2.021 A
2.022 - -
2.023 A
2.024 H^+
2.025 H_2O
2.026 H_3O^+
2.027 OH^-
2.028 H_3O^+
2.029 OH^-
2.030 H_3O^+

SELF TEST 3

3.01   b
3.02   e
3.03   d
3.04   a
3.05   c
3.06   Either order:
     a. fats
     b. oils
3.07   Either order:
     a. alkali
     b. antacid
3.08   laxative
3.09   hydroxide
3.010 physical
3.011 true
3.012 false

3.013 true
3.014 false
3.015 false
3.016 d
3.017 h
3.018 e
3.019 j
3.020 g
3.021 i
3.022 f
3.023 a
3.024 k
3.025 b

3.026 a. B
      b. N
      c. B
      d. B
      e. A
      f. B
      g. A
      h. A
      i. N salt is neutral; NaOH
         is a base
      j. B

SELF TEST 4

4.01 true
4.02 false
4.03 true

4.04 true

4.05 true

4.06 h

4.07 d

4.08 k

4.09 b

4.010 i

4.011 j

4.012 f

4.013 e
4.014 a
4.015 c
4.016 a

4.017 b

4.018 b

4.019 The splitting of a nucleus into smaller parts. Energy is released.

4.020 The joining of two nuclei into one. Energy is released.

4.021 a. 2
      b. 10
      c. 2
      d. 8
      e. 5

SELF TEST 1

1.01 m
1.02 l
1.03 a
1.04 n
1.05 i
1.06 k
1.07 h
1.08 b
1.09 j
1.010 d
1.011 e
1.012 o
1.013 g
1.014 f
1.015 c
1.016–1.023 Any order; any eight:
d, g, m, n, l, p, s, h, f
1.024–1.027 Any order:
s, b, o, c
1.028–1.030 Any order; any three:
n, l, d, r, m, a

1.031-1.032 Either order; any two:
e, a, f, h, l, n, s, i, d,
g, k, m, p
1.033-1.035 Any order; any three:
k, d, h, m, j, g, q, f, i,
s, c, l
1.036 a. process
b. breaking down (dissolving)
c. bloodstream (plasma)
1.037 saliva
1.038 5
1.039 2
1.040 3
1.041 1
1.042 4
1.043 See schematic of the digestive system in the preceding Science LIFEPAC. Check with the chart you have drawn.

SELF TEST 2

2.01 a
2.02 d
2.03 d
2.04 a
2.05 e
2.06 c
2.07 c
2.08 e
2.09 a
2.010 b
2.011 e
2.012 f
2.013 a
2.014 e
2.015 c

2.016 Any order:
a. vegetables
b. grains
c. dairy
d. protein
e. fruits
2.017 Either order:
a. acid (hydrochloric acid)
b. enzymes (pepsin)
2.018 carbohydrates (starches)
2.019 Either order:
a. gall bladder
b. pancreas
2.020 large intestine (colon)
2.021 4
2.022 3
2.023 5
2.024 2
2.025 1

2.026 e.g The grains group provides proteins, vitamins, and carbohydrates, therefore are most important. The vegetables and fruits groups provide vitamins, minerals, and some protein. The number of servings of grains, fruits, and vegetables provide all needed nutrients. Protein builds and repairs muscle.

2.027 teacher check for MyPlate servings content, variety, and economy of food items used.

SELF TEST 3

3.01 true

3.02 false

3.03 true

3.04 true

3.05 false

3.06 true

3.07 false

3.08 true

3.09 false

3.010 true

3.011 false

3.012 true

3.013 true

3.014 true

3.015 true

3.016 true

3.017-3.020 Any order:
3.017 a. calcium
b. milk, cheese, or green vegetables

3.018 a. phosphorus
b. milk, cheese, green vegetables, seafoods, cereals

3.019 a. iron
b. raisins, apricots, dried beans, peas, cereal, liver, eggs, meat, vegetables, beets

3.020 a. iodine
b. seafood, iodized salt

3.021-3.026 Any order:
3.021 a. Vitamin A
b. milk products, vegetable, egg yolks, liver

3.022 a. Vitamin D
b. fortified milk, fish, liver, sunshine, butter

3.023 a. Vitamin C
b. citrus fruits, raw fruits and vegetables

3.024 a. Vitamin B
b. milk products, whole grains, vegetables, peanuts, meat, fish, poultry, eggs, dried peas, dried beans

3.025 a. Vitamin E
b. vegetable oils, wheat germ, liver, lettuce

3.026 a. Vitamin K
b. green vegetables

3.027 a. grains
b. vegetables
c. fruits
d. dairy
e. protein
f. extras or discretionary calorie allowance

3.028 major dietary deficiency causing severe problems.

3.029 have minor dietary deficiencies with minor, limited symptoms.

3.030 infections, tissue hardness, partial to complete blindness, dry skin

3.031 eat foods containing carotene such as egg yolk, vegetables, yellow corn, or Vitamin A supplements.

3.032 a. related
b. a deficiency in others

3.033 fatigue, depression, irritability, low blood pressure, and loss of appetite.

3.034 yeast, rice bran, whole grain flour, meat, dried beans and peas, milk

3.035 physical weakness, easy bleeding of tissues, easy bruising, soft bones

3.036 Vitamin C

3.037 See Schematic of the digestive system in the preceding Science LIFEPAC. Check with the chart you have drawn.

SELF TEST 4

4.01 b
4.02 f
4.03 d
4.04 c
4.05 e
4.06 e
4.07 a
4.08 c
4.09 e
4.010 d
4.011 d
4.012 b
4.013 a
4.014 e or d
4.015 c
4.016 mouth (saliva glands)
4.017 esophagus

4.018 liver
4.019 gall bladder
4.020 stomach
4.021 pancreas
4.022 large intestine
4.023 small intestine
4.024 a. mouth: breaks up food, begins digestion of carbohydrates (starch)
b. stomach: liquifies food, digests starch, and begins digestion of protein
c. small intestine: digests protein and allows for absorption of nutrients through the walls into the bloodstream (plasma)
d. large intestine: collection of undigested solids, removal of excess water, and removal of waste.

4.025 the proper nutrient balance is necessary for proper cell reproduction and repair

4.026 Refined foods are generally less nutritious, less beneficial, and sometimes harmful due to overprocessing and the additives used to make the product more attractive and/or last longer.

4.027 Sample answers:
   a. carrots, corn, green beans, green and yellow vegetables, kale, broccoli
   b. bread, whole wheat cereal, rice, macaroni
   c. peaches, pears, citrus
   d. fish, poultry, beans, peas, peanuts, milk, cheese, cottage cheese

4.028 Sample answers:
   a. night blindness: deficiency of Vitamin A
   b. beriberi: Vitamin B complex deficiency
   c. scurvy: Vitamin C deficiency

4.029 teacher check for MyPlate components, variety in food type, and reasonable answer for the specific inclusion of each food.

4.030 teacher check for ideas which relate to the reverent attitude Romans 12:1,2 helps develop concerning the body as God's (Holy Spirit's) temple (dwelling place)

4.031 proper diet, proper exercise, cleanliness, proper rest, regular medical checkups.

SELF TEST 1

1.01	false		1.011	W
1.02	true		1.012	N
1.03	true		1.013	W
1.04	true		1.014	W
1.05	true		1.015	N
1.06	false		1.016	P
1.07	true		1.017	K
1.08	true		1.018	P
1.09	true		1.019	K
1.010	false		1.020	K

SELF TEST 2

2.01   energy

2.02   fusion

2.03   proton

2.04   kinetic

2.05   reflected

2.06   work

2.07   oxidation

2.08   friction

2.09   conduction

2.010   thermonuclear

2.011   Atomic nuclei of the same element are shot at each other at high speed and high heat.

2.012   The earth absorbs the sun's radiation and warms the air by convection.

2.013   The air is heated, expands, and rises. Cool air falls and is heated in turn. Slowly the room is heated by the warming and cooling air.

2.014   Force and distance must be in the same direction. If the force is up and the suitcase moves forward, no work is done.

2.015   true

2.016   false

2.017   false

2.018   true

2.019   false

SELF TEST 3

3.01	b		3.013	true
3.02	a		3.014	false
3.03	a		3.015	true
3.04	c		3.016	true
3.05	b		3.017	false
3.06	a		3.018	true
3.07	c		3.019	false
3.08	a		3.020	false
3.09	c		3.021	true
3.010	b		3.022	false

3.011 a. mechanical (flipping the switch)

      b. electrical

      c. radiant (the light from the bulb)

      d. heat

3.012 a. mechanical (bouncing the ball)

      b. mechanical (the ball moves)

      c. mechanical (the glass breaks)

3.023	false
3.024	true
3.025	true
3.026	true

SELF TEST 1

1.01   e

1.02   c

1.03   a

1.04   g

1.05   d

1.06   b

1.07   j

1.08   i

1.09   h

1.010 f

1.011

1.012

1.013– 1.015   Any order:

1.013   strength of current

1.014   number of turns in coil

1.015   material in the core

1.016–1.017   Either order:

1.016   dropped

1.017   hammered

1.018   The earth is a big bar magnet in its magnetic behavior.

1.019   An electromagnet is made by looping wire around an iron core and attaching it to a source of electricity.

1.020   A magnetic compass is not useful for navigating a submarine because the submarine is made of steel.

SELF TEST 2

2.01   g

2.02   f

2.03   c

2.04   k

2.05   d

2.06   i

2.07   a

2.08   e

2.09   h

2.010 b

2.011 b

2.012 a

2.013 c

2.014 d

2.015 d

2.016 magnet

2.017 static electricity

2.018 inverse square law

2.019 circuit

2.020 insulators or nonconductor

2.021 ampere

2.022 alternating current

2.023 Davy

2.024 power

2.025 alnico

2.026 Any order:

a. like charges repel

b. unlike charges attract

c. charged objects attract uncharged objects

2.027 magnetic force decreases as distance from a pole increases

2.028 six volts

2.029 nine volts

2.030 87.6 watts

SELF TEST 3

3.01 d

3.02 g

3.03 a

3.04 e

3.05 b

3.06 h

3.07 c

3.08 f

3.09 j

3.010 m

3.011 o

3.012 k

3.013 i

3.014 l

3.015 n

3.016 true

3.017 false

3.018 false

3.019 false

3.020 true

3.021 true

3.022 false

3.023 true

3.024 true

3.025 true

3.026 c

3.027 b

3.028 a

3.029 a

3.030 d

3.031 a

3.032 c

3.033 b

3.034 b

3.035 a

3.036 both break the circuit

3.037 stationary charge

3.038 Examples; Any order:

a. clean

b. can be used in isolated areas

c. unending supply

3.039 Examples; Any order:

a. clean

b. short transmission lines

c. serve many people

3.040 110 volts

3.041 1,725 watts

3.042 880 volts

SELF TEST 1

1.01   d

1.02   d

1.03   a

1.04   b

1.05   c

1.06   100

1.07   1

1.08   2,500

1.09   73,000

1.010   0.147

1.011   1,000 km (measurements
          will vary from 980-1100 km)

1.012

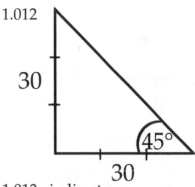

1.013   indirect

1.014   geometry

1.015   kilometer

1.016   10

1.017   meter

1.018   estimation

SELF TEST 2

2.01   b

2.02   a

2.03   c

2.04   c

2.05   a

2.06   c

2.07   b

2.08   c

2.09   a or b

          Note on 2.09: The satellite
          experiences little or no resistance
          force of friction; but it constantly
          experiences the attractive
          gravitational force, which pulls
          it into a circular path.

2.010   b

2.011   vector

2.012   force

2.013   gravity

2.014   size or magnitude

2.015   Italy

2.016   earth

2.017   England

2.018   metric

2.019   metric

2.020   indirect

2.021   true

2.022   false

2.023   true

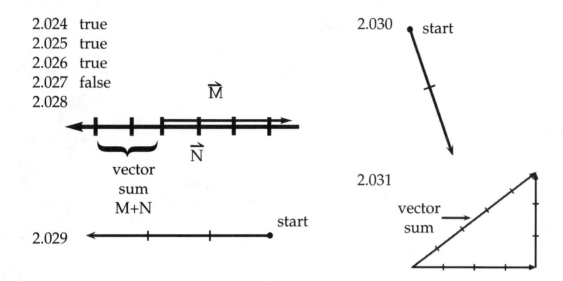

2.024 true
2.025 true
2.026 true
2.027 false
2.028

2.029

2.030 start

2.031 vector sum

2.032 a. approximately 43 yards
       b. indirect

SELF TEST 3

3.01   250
3.02   1.04
3.03   3,700
3.04   7,000
3.05   .091
3.06   newton
3.07   meter
3.08   joule
3.09   watt
3.010 joule
3.011 second
3.012 P
3.013 P
3.014 B
3.015 K
3.016 B

3.017 false
3.018 true
3.019 true
3.020 true
3.021 true
3.022 false
3.023 true
3.024 true
3.025 false
3.026 false
3.027 energy
3.028 distance
3.029 potential
3.030 potential
3.031 kinetic
3.032 joule

3.033 power

3.034 Galileo

3.035 second

3.036 third

3.037 No elevational change is mentioned. Therefore, no work is done because the distance vector (horizontal) is not in the same direction as the force vector (vertical).

3.038 work = force x distance

   work = 6 N x 3 m = 18 J

3.039 work = force x distance

   work = 2,500 lbs. x 16 ft.

   = 40,000 ft. lbs.

3.040 80 m

3.041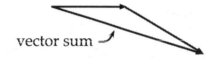

vector sum

SELF TEST 1

1.01   true

1.02   true

1.03   true

1.04   true

1.05   false

1.06   false

1.07   true

1.08   b

1.09   d

1.010   Examples:

    a.   shoes on sidewalk

    b.   brakes

    c.   tires on the road

1.011   Friction decreases the efficiency of all machines.

1.012   Any order:

    a.   lubrication

    b.   shape

    c.   material

1.013– 1.015   Examples:

1.013   a.   resistance to motion when one object is pulled over a surface

    b.   any flat object moving over a flat surface

1.014   a.   resistance of a round surface to another surface

    b.   ball bearings, wheels, log rollers

1.015   a.   resistance of a surface to movement through gas or liquid

    b.   cars or plane to air, boats to air or water, submarines to water

1.016   a. $\mu = \dfrac{F_f}{F_n} = \dfrac{6}{18} = \dfrac{1}{3} = 0.\overline{33}$

    b $\mu = \dfrac{F_f}{F_n} = \dfrac{3}{18} = \dfrac{1}{6} = 0.1\overline{6}$

SELF TEST 2

2.01	true
2.02	false
2.03	true
2.04	true
2.05	true
2.06	c
2.07	a
2.08	e
2.09	f
2.010	b
2.011	b
2.012	a

2.013 b

2.014 c

2.015 a

2.016 $\mu = \dfrac{F_f}{F_n} = \dfrac{2N}{8N} = \dfrac{1}{4} = 0.25$

2.017 a. $\dfrac{1m \times 1{,}200\ N}{3m} = 400\ N$

      b. 450 N

      c. $\dfrac{3m}{1m} = 3$

      d. $\dfrac{1{,}200N}{450\ N} = 2.\overline{6}$

      e. $\dfrac{2.6}{3} = 0.8\overline{6}$ or 87%

SELF TEST 3

3.01	false
3.02	true
3.03	true
3.04	false
3.05	true
3.06	true
3.07	true
3.08	true
3.09	resistance
3.010	gears
3.011	friction
3.012	effort force
3.013	axle
3.014	$\dfrac{18}{3} = 6$

3.015

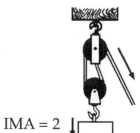

IMA = 2

3.016 $\dfrac{39}{13} = 3$

      $4\,^1/_2 \times 3 = 13\,^1/_2$ times

3.017 a. IMA $= \dfrac{10m}{2m} = 5$

      b. AMA $= \dfrac{5N}{2N} = 2\,^1/_2$

      c. efficiency $= \dfrac{AMA}{IMA} \times 100\%$

         $= \dfrac{2\,^1/_2}{5} \times 100\%$

         $= 50\%$

3.018 Work is always lost in overcoming friction.

SELF TEST 4

4.01   g

4.02   j

4.03   a

4.04   h

4.05   f

4.06   b

4.07   d

4.08   e

4.09   i

4.010 k

4.011 friction

4.012 one

4.013 effort force

4.014 direction

4.015 length

4.016 lubricants

4.017 thinner

4.018 $\mu = \dfrac{f_f}{f_n} = \dfrac{16N}{24N} = 0.6\overline{6} = 0.67$

4.019 a. $6/4 = 1.5$

      b. $\dfrac{9,000}{7,500} = 1.2$

      c. $1.2/1.5 = .8$ or 80%

4.020 $IMA = \dfrac{9cm}{3cm} = 3$

4.021 $24/8 = 3$

4.022 $120/4 = 30$ newtons

SELF TEST 1

1.01	false	1.024	a
1.02	true	1.025	d
1.03	false	1.026	b
1.04	true	1.027	a
1.05	false	1.028	c
1.06	false	1.029	d
1.07	false	1.030	c

1.08 true

1.031 a. $6 CO_2$

b. $12 H_2O$

c. light

$\longrightarrow$

chlorophyll

d. $C_6H_{12}O_6$

e. $6 O_2$

f. $6 H_2O$

1.09 true

1.010 true

1.011 g

1.012 i

1.013 k

1.032 Any order:

a. machines

b. chemicals

c. hybrid plants

1.014 c

1.033 Any order:

1.015 d

a. light

1.016 j

b. water

c. nitrogen

1.017 h

d. phosphorus

e. potassium

1.018 f

f. hydrogen

g. oxygen

1.019 a

h. carbon

i. trace minerals

1.020 b

1.021 a

1.034 Examples:

a. root - carrot

b. fruit - apple

1.022 d

c. stem - asparagus

d. leaf - cabbage

1.023 b

e. flower - cauliflower

SELF TEST 2

2.01	false
2.02	true
2.03	false
2.04	false
2.05	false
2.06	true
2.07	true
2.08	true
2.09	false
2.010	true
2.011	g
2.012	j
2.013	h
2.014	k
2.015	e
2.016	a
2.017	b
2.018	c
2.019	f
2.020	i or d
2.021	d
2.022	d

2.023 a
2.024 a
2.025 c
2.026 c
2.027 c
2.028 a
2.029 a
2.030 a
2.031 Any order:
    a. clover
    b. peas
    c. beans
    d. alfalfa or peanuts
2.032 The bacteria provide the plant with nitrogen compounds and legumes provide the bacteria with food.
2.033 Either order:
    a. Decomposition releases important elements back into their cycles.
    b. Decomposition rids the earth of excessive organic material.
2.034 Examples:
    a. drying– dried fruits
    b. salting– beef jerky, ham
    c. canning– peas
    d. freezing– strawberries
    e. refrigerating– lettuce

SELF TEST 3

3.01	true	3.05	false
3.02	true	3.06	false
3.03	false	3.07	false
3.04	true	3.08	true

3.09   false

3.010 true

3.011 ecology

3.012 l

3.013 hybrid

3.014 energy or biomass

3.015 legume

3.016 crop rotation

3.017 60

3.018 nitrogen

3.019 scavenger

3.020 carbon dioxide

3.021 e

3.022 h

3.023 g

3.024 k

3.025 c

3.026 d

3.027 a

3.028 j

3.029 b

3.030 f

3.031 c

3.032 c

3.033 d

3.034 b

3.035 b

3.036 c

3.037 c

3.038 a

3.039 a

3.040 c

3.041 study of organisms in relation
       to their environment

3.042 place where an organism lives

3.043 a plant that provides food for a
       community

3.044 an animal that eats plants or
       other animals

3.045 all organisms living in a certain area

3.046 an animal that eats only other animals

3.047 an animal that eats only plants

3.048 a condition that limits the
       numbers of an organism

3.049 an animal that eats plants and animals

3.050 an animal that eats dead material

3.051 Examples; Any order:

    a. soil – proper plowing

    b. forests – recycle paper

    c. minerals – recycle minerals

    d. fossil fuel – don't use disposable
       plastics

    e. air – walk instead of ride

    f. water – shorter showers

    g. wildlife– careful with chemical sprays

    h. wilderness areas – find way to
       pay for more wilderness areas

3.052 Examples; Any order:

    a. greed – kill rare animals

    b. population – trash

    c. ignorance – kill natural controls

    d. technology — chemicals pollute
       air and water

SELF TEST 1

1.01　g

1.02　k

1.03　a

1.04　i

1.05　h

1.06　b

1.07　c

1.08　j

1.09　e

1.010　d

1.011　false

1.012　true

1.013　true

1.014　false

1.015　false

1.016　true

1.017　true

1.018　Any order:

    a.　observation

    b.　investigation

    c.　experimentation

1.019　a.　meter

    b.　liter

    c.　kilogram

1.020　Either order:

    a.　hydrogen

    b.　oxygen

1.021　Any order:

    a.　physical

    b.　chemical

    c.　nuclear

1.022　a.　Niels Bohr

    b.　orbits

1.023　nuclear fission

1.024　more than 5 million

1.025　buoyancy

1.026　chemical equation

1.027　chemical formula

1.028　Metrics uses a uniform scale of decimals. All units are in multiples of ten and have uniform names.

1.029　Any five, any order:
color, taste, odor, hardness, brittleness
or luster, form, density, buoyancy

1.030　Water under pressure boils at a higher temperature, thus cooking the food in less time.

SELF TEST 2

2.01	f	2.015	a
2.02	h	2.016	Any order:
			a. force
2.03	g		b. speed
2.04	a		c. change direction of force
2.05	j	2.017	friction
		2.018	a. actual mechanical advantage
2.06	b		b. ideal mechanical advantage
2.07	i	2.019	a. output
			b. input
2.08	d	2.020	Either order:
2.09	e		a. appearance
			b. composition
2.010	c	2.021	a metal made by melting together two or more metals
2.011	b		
2.012	c	2.022	oil, grease, etc. for putting on machines to overcome friction
2.013	b	2.023	method of transferring heat by moving molecules rapidly in circular currents; forms winds
2.014	d		

SELF TEST 3

3.01	Either order:	3.09	false
	a. physical		
	b. chemical	3.010	true
3.02	Any order:	3.011	true
	a. proteins		
	b. minerals	3.012	f
	c. fats and oils		
	d. carbohydrates	3.013	i
	e. vitamins		
	f. water	3.014	g
3.03	alimentary	3.015	a
3.04	true	3.016	c
3.05	false	3.017	h
3.06	false	3.018	e
3.07	true	3.019	d
3.08	true		

3.020 takes in fuel; fuel is changed chemically and physically to release energy and accomplish something; discharges waste

3.021 by developing bad habits; not eating healthful foods; not practicing good hygiene

3.022 Evaporation, Storm Clouds, Precipitation (rain, snow...), Earth

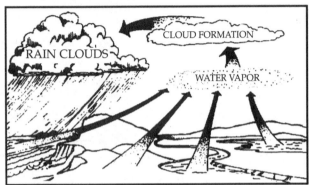

THE WATER CYCLE

3.023 Conduction: heat is transferred directly from one molecule to the next
Convection: heat is transferred by air currents
Radiation: heat travels in waves through the air

3.024 An electric current is passed through water. Positive ions are attracted to the negative pole and negative ions to the positive pole. Gas collects in the test tubes.

SELF TEST 4

4.01 d
4.02 i
4.03 h
4.04 g
4.05 c
4.06 j
4.07 b
4.08 e
4.09 f
4.010 a
4.011 Either order:
    a. likes (interests)
    b. dislikes (abilities)
4.012 résumé
4.013 a. potential energy
    b. kinetic energy
4.014 repel
4.015 social security card
4.016 Any five; any order:
    a. dependability
    b. having initiative
    c. honesty

    d. courtesy
    e. good health habits
    or efficiency, neatness, pleasantness, friendliness
4.017 And also that every man should eat and drink, and enjoy the good of all his labor, it is the gift of God.
4.018 Example:
If I get a job in a science-related field I will have to use metrics. The United States is converting to the metric system so I will have to use it in everyday life.
4.019 A mixture is not changed chemically; it can be separated into its original components. A compound has undergone a chemical change and cannot easily be separated into its original elements.
4.020 a

4.021 a

4.022 b

4.023 b

4.024 a

4.025 b

4.026 b

4.027 a

4.028 c

4.029 e

4.030 f

4.031 b

4.032 g (b)

4.033 c

4.034 a

4.035 d

4.036 b

4.037 a

Notes

146

Notes

1. false
2. true
3. true
4. false
5. true
6. false
7. true
8. false
9. false
10. true
11. 620,000
12. 1,000
13. 11.1
14. e
15. h
16. b
17. d
18. c
19. g
20. a
21. God's Word
22. Darwin
23. solar energy
24. inclined plane
25. Galileo
26. Any three; any order:
good medicine, life-support machines, synthetic foods, improved food supply, comforts, conveniences
27. Many ancient writings were lost.
28. Example; Any order:
a. pollution—produced by industry
b. food shortages from increased population
c. possible harm from synthetic foods

1.	d	25.	b.	liquid
2.	e	26.	O	
3.	k	27.	X	
4.	f	28.	O	
5.	l	29.	O	
6.	h	30.	X	
7.	c	31.	O	
8.	a	32.	X	
9.	i	33.	X	
10.	b	34.	O	
11.	true	35.	X	
12.	true	36.	X	
13.	false	37.	23 or 22.991	
14.	true	38.	11	
15.	false	39.	12	
16.	false	40.	Na	
17.	true	41.	11	
18.	true	42.	8	
19.	false	43.	4	
20.	true	44.	5	
21.	a. salt	45.	3	
22.	b. atomic number	46.	2	
23.	a. protons plus neutrons	47.	3	
24.	c. 103			

1. l
2. j
3. i
4. d
5. f
6. h
7. k
8. m
9. g
10. b
11. c
12. e
13. true
14. true
15. true
16. false
17. a.
18. c.
19. A physical change is a change in appearance with no change in composition.
20. A chemical change produces a new substance with a different molecular structure than the original.
21. A nuclear change is a change involving the nucleus of the atom.

22. Neutralization is the combination of an acid and a base to produce water and salt.
23. a. A
    b. B
    c. B
    d. A
    e. A
    f. B
    g. B
    h. N
    i. B
    j. B
24. Any three; any order:
    a. melting of ice
    b. fertilizer
    c. photography or table salt copper plating, etc.
25. Any three; any order:
    a. bleaches
    b. batteries
    c. paints
       or
       fertilizers
       explosives
       preservatives, etc.

1. b
2. d
3. c
4. a
5. d
6. c
7. d
8. c or d
9. b
10. a
11. c
12. e
13. d
14. e
15. e
16. Any order:
    a. grains
    b. vegetables
    c. fruits
    d. dairy
    e. protein
17. Any order:
    a. carbohydrates
    b. fats
    c. proteins
    d. minerals
    e. vitamins
    f. water
18. Examples:
    a. beriberi: vitamin B com-
          plex deficiency
    b. scurvy: vitamin C
          deficiency
    c. night blindness: vitamin A
          deficiency
19. b
20. c
21. d

22. a
23. b
24. a. liver
    b. gall bladder
    c. stomach
    d. large intestine
    e. small intestine
25. Any order:
    a. proper diet
    b. proper exercise
    c. cleanliness
    d. regular checkups
    e. proper rest
26. Good nutrition prevents nutritional deficiency diseases which have their symptoms in poor hair, skin, and teeth. Proper diets are essential to cell reproduction and repair, they fight against infections, and provide the necessary energy to be mentally and physically alert.
27. The menus will be different but check for MyPlate serving, content, variety, and economy of food items used.
28. The specific examples will vary. Look for examples that illustrate the students specific presentation. The main idea to look for is: God's natural food is best because it was specifically designed to fill our bodily needs. Man's synthetic products and attempts to make food more appealing and/or last longer have introduced chemicals that can be very harmful to the delicate human body.

1. true
2. true
3. true
4. false
5. true
6. true
7. true
8. true
9. true
10. false
11. work
12. electrons
13. fusion
14. conduction
15. burning
16. c
17. a
18. d
19. f
20. e

21. i
22. b
23. g
24. j
25. h
26. Conduction transfers heat by direct contact of one molecule with another, vibrating against each other. Convection is heating by physical circulation in a fluid due to a difference in density due to heating and cooling within the fluid.
27. Entropy describes the tendency of the universe to "run down" and become more disordered. To a Christian this simply proves the presence of God, the Creator and sustainer of the universe, the same yesterday, today, and tomorrow.

1.	g	25.	true
2.	f	26.	d
3.	e	27.	d
4.	h	28.	a
5.	d	29.	d
6.	a	30.	b
7.	j	31.	b
8.	k	32.	c
9.	c	33.	c
10.	b	34.	a
11.	true	35.	d
12.	false	36.	current
13.	true	37.	William Gilbert
14.	true	38.	Benjamin Franklin
15.	true	39.	conductors
16.	true	40.	magnetic
17.	false	41.	reactors
18.	true	42.	series
19.	true	43.	circuit
20.	true	44.	electromagnet
21.	false	45.	Ohm
22.	false	46.	220 volts
23.	true	47.	sixteen watts
24.	false		

1. d
2. e
3. f
4. b
5. f
6. a
7. g
8. true
9. true
10. false
11. true
12. true
13. Newton
14. Galileo
15. vector
16. first
17. second
18. third

19. third
20. second
21. kinetic
22. power
23. 50 km
24.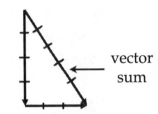

vector sum

25. $$\text{power} = \frac{\text{force x distance}}{\text{time}}$$

$$\text{power} = \frac{10\,N \times 8\,m}{4\,\text{sec.}} = 20\text{ watts} = 20\text{ joules/sec.}$$

1. true
2. true
3. true
4. true
5. true
6. false
7. true
8. true
9. false
10. true
11. c
12. c
13. a

14. b
15. b
16.
   a. $\dfrac{12 \text{ m}}{4 \text{ m}} = 3$

   b. $\dfrac{1{,}200 \text{ N}}{400 \text{ N}} = 3$

   c. $\dfrac{3}{3} = 100\%$

17. B

18. $\mu = \dfrac{f_f}{f_n} = \dfrac{8 \text{ N}}{20 \text{N}} = \dfrac{2}{5}$ or 0.4

1.	f		22.	a
2.	i		23.	a
3.	c		24.	a
4.	e		25.	c
5.	a		26.	b
6.	h		27.	d
7.	d		28.	a
8.	k		29.	d
9.	g		30.	a
10.	b		31.	f
11.	true		32.	k
12.	true		33.	i
13.	false		34.	c
14.	true		35.	a
15.	true		36.	j
16.	false		37.	e
17.	true		38.	d
18.	true		39.	g
19.	true		40.	b
20.	true			
21.	a			

1.	e	30.	a	
2.	i	31.	g	
3.	h	32.	c	
4.	f	33.	e	
5.	a	34.	a	
6.	c	35.	f	
7.	b	36.	a	
8.	k	37.	b	
9.	g	38.	Either order:	
10.	d		a. adaptability (dependability)	
11.	true		b. resiliency (efficiency)	
12.	true	39.	Any order:	
13.	false		a. observation	
14.	false		b. investigation	
15.	true		c. experimentation	
16.	false	40.	a. meter	
17.	true		b. kilogram	
18.	false		c. liter	
19.	true	41.	harder	
20.	false	42.	Either order:	
21.	a		a. protons	
22.	b		b. neutrons	
23.	a	43.	a– b Either order:	
24.	a		a. hydrogen	
25.	b		b. oxygen	
26.	a		c. electrolysis	
27.	c	44.	atomic bomb	
28.	b	45.	heat (friction)	
29.	b			

1. true
2. true
3. true
4. false
5. true
6. true
7. false
8. true
9. false
10. true
11. 534
12. 1,000
13. 19.7
14. h
15. f
16. a
17. e
18. b
19. d
20. c
21. God
22. shaduf
23. solar energy
24. Einstein
25. Copernicus
26. Examples; any order:
    a. food shortage
    b. fuel shortage
    c. transportation
27. Example:
    The Bible was read more widely
    by people in their homes.
    Later other books were printed.
28. Examples; any order:
    a. space exploration
    b. communication
    c. medical advances

1.	c		41.	6
2.	j		42.	3
3.	i		43.	3
4.	f		44.	2
5.	b		45.	6
6.	k		46.	7
7.	e			
8.	a			
9.	d			
10.	h			
11.	false			
12.	true			
13.	true			
14.	false			
15.	false			
16.	true			
17.	true			
18.	false			
19.	false			
20.	false			
21.	b. sodium			
22.	a. protons and neutrons			
23.	c. gas			
24.	c. electrons			
25.	c. 100			
26.	x			
27.	o			
28.	x			
29.	x			
30.	o			
31.	x			
32.	o			
33.	o			
34.	x			
35.	x			
36.	12			
37	6			
38.	6			
39.	C			
40.	6			

1. change from liquid to gas (vapor)
2. change from gas (vapor) to liquid
3. involves both evaporation and condensation
4. i
5. h
6. k
7. d
8. j
9. g
10. a
11. b
12. c
13. e
14. false
15. true
16. false
17. true
18. true
19. a. A
    b. B
    c. B
    d. N
    e. A
    f. B
    g. B
    h. A
20. Examples; any order:
    a. melting of ice
    b. fertilizer
    c. photography
       or copper plating
       gun powder
       table salt
       preservatives
21. Examples; any order:
    a. bleaches
    b. batteries
    c. paints
       or fertilizers
       explosives
       preservatives

1. true
2. false
3. true
4. true
5. false
6. true
7. true
8. false
9. false
10. true
11. Any order:
    a. calcium
    b. phosphorus
    c. iron
    d. iodine
12. Any order:
    a. grains
    b. vegetables
    c. dairy
    d. protein
    e. fruits
13. Any order:
    a. carbohydrates
    b. fats
    c. proteins
    d. minerals
    e. vitamins
    f. water
14. f
15. m
16. j
17. i
18. c
19. e
20. n
21. k
22. l
23. d
24. b
25. o
26. a
27. h

28. mouth
29. stomach
30. small intestine
31. antiseptic
32. cleanliness, sanitation

1. true
2. false
3. true
4. true
5. false
6. true
7. true
8. true
9. true
10. false
11. b
12. d
13. a
14. b
15. a
16. d
17. e
18. d
19. k
20. i
21. a
22. h
23. c
24. f
25. b
26. j
27. Example:
    Kinetic energy is energy of
    motion. Potential energy is
    stored energy.
28. Example:
    Force is a push or a pull.
    Work is force moving through
    a distance.

1.	g	22.	c
2.	a	23.	d
3.	f	24.	a
4.	c	25.	d
5.	e	26.	true
6.	b	27.	false
7.	m	28.	true
8.	i	29.	true
9.	k	30.	false
10.	l	31.	false
11.	h	32.	false
12.	d	33.	true
13.	c	34.	true
14.	a	35.	true
15.	c	36.	false
16.	a	37.	true
17.	c	38.	true
18.	a	39.	false
19.	b	40.	true
20.	d	41.	110 volts
21.	a	42.	770 watts = 770 joules/sec.

1. c
2. h
3. a
4. d
5. f
6. e
7. b
8. false
9. true
10. true
11. true
12. false
13. true
14. estimate
15. the French
16. direction
17. Newton
18. joules
19. motion
20. Newton
21. size or magnitude
22. scalar
23. 70 km
24.

25. power   = $\dfrac{\text{force x distance}}{\text{time}}$

power   = $\dfrac{20 \text{ N} \times 10 \text{ m}}{5 \text{ sec.}}$ = 40 joules/sec. = 40 watts

Science 808 Alternate Test Key

1. true
2. true
3. true
4. true
5. true
6. false
7. false
8. true
9. true
10. true
11. c
12. a
13. b
14. d
15. a
16. a. 2
    b. 1.5
    c. 75.0%
17. A
18. 0.25

1.	g	21.	a
2.	i	22.	d
3.	j	23.	c
4.	k	24.	c
5.	h	25.	d
6.	a	26.	a
7.	b	27.	a
8.	e	28.	d
9.	d	29.	b
10.	c	30.	a
11.	true	31.	g
12.	false	32.	j
13.	true	33.	i
14.	true	34.	f
15.	true	35.	k
16.	true	36.	a
17.	true	37.	e
18.	false	38.	c
19.	false	39.	b
20.	true	40.	d

1. e
2. i
3. r
4. m
5. j
6. u
7. a
8. t
9. k
10. b
11. q
12. f
13. h
14. c
15. d
16. l
17. g
18. s
20. o
21. p
22. c
23. d
24. b
25. e
26. true

27. true
28. false
29. true
30. false
31. true
32. false
33. false
34. true
35. false
36. false
37. false
38. true
39. false
40. true
41. false
42. true
43. moving
44. density
45. Either order:
 a. created
 b. destroyed
46. Any two:
 a. solar
 b. geothermal or nuclear
47. eat

48. Examples:

Magnetism	Electricity
a. Like poles repel.	Like charges repel.
b. Unlike poles attract.	Unlike charges attract.
c. Magnet has field of force.	Wire carrying electricity has field.
or Magnet through coil makes electricity flow.	Wire carrying current magnetizes iron core.

49. movement of molecules

50. Examples:
No definitive answer. Students may take one of two attitudes.
1. Everyone will stockpile and use weapons to destroy the world.
2. Every nation will fear the strength of every other nation and no one will use weapons.